LIFELINE EARTHQUAKE ENGINEERING:

Performance, Design and Construction

Proceedings of a Symposium
sponsored by the
Technical Council on Lifeline Earthquake Engineering
of the
American Society of Civil Engineers
in conjunction with the
ASCE National Convention
San Francisco, California
October 4-5, 1984

James D. Cooper, Editor

Published by the
American Society of Civil Engineers
345 East 47th Street
New York, New York 10017-2398

The Society is not responsible for any statements made or opinions expressed in its publications.

Library of Congress Catalog Card No: 84-71890
ISBN 0-87262-414-5
Manufactured in the United States of America.

FOREWORD

A major effort of the Technical Council on Lifeline Earthquake Engineering culminated with the publication of "Advisory Notes on Lifeline Earthquake Engineering" in October 1983. The Notes provide a review of the current state-of-the-art in the seismic design practice for lifeline facililties including: Electric Power and Telecommunications Systems; Oil and Gas Pipeline Systems; Transportation Lifelines; and Water and Sewage Lifelines. In addition to providing some insight into the performance of these lifelines in past earthquakes, the Notes refer the reader to available codes of practice or specifications and present design approaches to be taken in the seismic design process.

Papers published in these Proceedings were presented at the four session symposium entitled *Lifeline Earthquake Engineering: Performance, Design and Construction* which was held in conjunction with the 1984 ASCE Annual Convention in San Francisco, California. This text is intended to be a practical follow-up to the Advisory Notes. Papers included herein highlight state-of-the-practice step-by-step procedures for the seismic design of transportation, water, sewer, power, and communications structures. Case histories are discussed.

Each of the papers included in the proceedings has received two positive peer reviews. All papers are eligible for discussion in the Journal of Technical Topics in Civil Engineering. All papers are eligible for ASCE awards.

Roy A. Imbsen, Chairman of the TCLEE Transportation Lifeline Committee chaired the convention session on Bridges and Tunnels. Anshel J. Schiff chaired the session on Risk, Retrofit and Transportation, and Duane B. Ford Chairman of the TCLEE Water and Sewage Committee chaired the session on Water, Sewer, Power and Communications. A fourth panel discussion session entitled Issues at the Interfaces of Lifeline Earthquake Engineering was chaired by Robert A. Olson. Panel members included Karl V. Steinbrugge, Richard Andrews, Terence Haney, Charles C. Thiel, Jr., and Richard S. Olson. No formal papers were prepared for that session.

James D. Cooper
Editor

CONTENTS

Section I—Bridges and Tunnels

Section II—Risk Retrofit and Transportation

Section III—Water, Sewer, Power and Communications

Engineering Liabilities for Design of a Bridge That Failed During the San Fernando Earthquake

By
Irwin Schulman*

FOREWORD

On February 9, 1971, the San Fernando Valley in Southern California experienced a major earthquake. Major damage to structures near the epicenter resulted in numerous deaths. Among the victims were two men who were killed when a bridge connector ramp of a newly constructed freeway interchange collapsed. The heirs sued the State of California, alleging that the highway was in a dangerous condition because the seismic design provisions in effect for bridges at the time when the bridge was designed failed to incorporate the most recent aseismic design practices.

The State defended itself by presenting substantial evidence that the structure was built in accordance with reasonably approved engneering plans and standards. The major issues concerned the earthquake design criteria selected by the State engineers. The State successfully defended the use of these design criteria.

VIS MAJOR

Research into the laws, both legislative and those developed through the court system concerning engineers' liability for earthquake damage, have turned up few materials. It is a generally held belief that man is not responsible for "an Act of God". That term has been interchangeable with the words "irresistible superhuman cause."

An early California statute(1) dealing with acts of God was one relating to liability for loss of the property of guests placed under the care of an innkeeper. That statute excepted losses "occasioned by an irresistible superhuman cause." An early California Supreme Court decision(2) interpreted that statute in the following terms:

*Attorney for State of California, Department of Transportation, 120 South Spring Street, Suite 500, Los Angeles, California 90012

> "The words 'irresistible superhuman cause' are equivalent in meaning to the phrase 'the Act of God', and refer to those natural causes the effects of which cannot be prevented by the exercise of prudence, diligence, and care, and the use of those appliances which the situation of the party renders it reasonable that he should explore."

It differentiated between losses occasioned by an accidentally caused fire (liability) and one caused by lightning (no liability). The sophistication of the engineering behind the design of the lightning arrestor or rod was not an issue.

A variety of other cases have dealt with the "Act of God". It has been said that it is an accident against which ordinary skill is not expected to provide.(3) It has also been intermixed with the terms "vis major" which a law dictionary(4) defines as "a higher force: an irresistible force. An event which cannot be definitely foreseen or controlled. The test of 'vis major' is whether under the particular circumstances there was such an insuperable interference without the party's intervention as could not have been prevented by the exercise of prudence, diligence, and care. . . ."

The courts have held earthquakes to be an act of God.(5) However, is all damage caused by an earthquake excusable as caused by an act of God? If an earthquake causes damage to only one house many miles away and no damage to those near its epicenter can the damage be excused as being caused by an irresistible force? If a severe rainstorm causes widespread damage to many properties should it be an excuse that it was an act of God even though it was of a predictable intensity? The answer to both of those simplified questions is in the negative.

Unfortunately, the facts underlying most occurrences are not as simple as a hypothetical. These are complex questions of fact. First, there is the intensity of the occurrence or force; second, there is the criteria utilized in the design of an item to enable it to withstand reasonably expected forces of nature.

DEFENSE ASPECTS OF THE LAWSUIT

Defenses included the act of God defense and the reasonableness of the governmental decisions in design. Most defenses are questions for the jury. The "design defense" is a question to be decided by the court without the intervention of a jury.(6)

There were no California cases explaining or shedding light on the earthquake problem. The lawsuit which arose from the Sylmar earthquake bridge collapse was not decided on the appellate level on the factors of the design.(7) The design defense by the State was presented on the legal assumption that the damages to the plaintiffs were design caused but that the design had a reasonable basis for adoption. If the ruling on that issue had been in the negative, the case would have proceeded to a jury trial where the plaintiff would have had to prove that the deaths were caused by the design of the bridge and the State would have been liable unless it could have proved that the bridge would have collapsed even if the design had not been defective or the decision to cut down on its earthquake collapse-resisting features was reasonable because of cost and/or competing projects which required the funds.

A large percentage of court decisions have involved insurance. In determining whether two causes, one excluded (act of God) and one included as an insured risk (water damage), acting together to produce the damage, the courts have used terms such as "efficient cause", "intervening, but not superceding cause", "prime or moving cause", etc. In those types of cases it has not been found that, but for the condition of the subject, the force would not have caused the damage. Recovery can be had even though a diseased or infirm condition appears to actually contribute to the cause of death, or the accident sets in progess the chain of events leading to the death, or if it is the prime or movng cause. Brooks v. Metropolitan Life (1945) 27 Cal.2d 305.

THE BRIDGE

Numerous design features were pointed to by the plaintiffs to establish that the bridge was not designed in accordance with reasonable design criteria. The bridge which was involved in this action was the connector from Route 210 to Route 5. It was 770 feet in length with nine piers, seven spans, and a "hinge" in span 3.

The design features criticized were the following:

1. The hinge expansion joint was unrestrained and not equipped to accept a larger displacement than 14 inches;

2. The columns in the middle of the bridge had splices which violated the AASHTO Code and the Code of the American Concrete Institute (ACI);

3. The seismic formula should have been the same as is in the SEAOC Code which provided for "C" to be a minimum of .10 for one story buildings, the Bridge Department

which lifted its seismic formula from that code provided that "C" should be a maximum of .10;

4. There was a criticism to the use of a "K" factor of .67 and not following structural steel reinforcement specifications called for when .67 is used for the "K" factor;

5. The imbedment of the #18 bars into the piles was of an insufficient depth;

6. No provision had been made to prevent diagonal tension in the footings; and

7. The unrestrained hinge with only a 14-inch seat would have unseated itself in a lesser earthquake.

THE TRIAL

The trial of the design defense took seven weeks and resulted in a judgment for the State. The court determined its function as the determination of the hearing as to whether or not there was substantial evidence justifying the adoption of the design of the bridge which was the gravamen of the consolidated lawsuits.

The Court considered the following evidence. One engineer designed 15 bridges in this interchange project. The 210/5 was the only one destroyed by the earthquake, although all suffered some damage.

The critical bridge was 770 feet long with seven spans and a "hinge" in span 3. The fact that this key had no design factor to restrain horizontal movement during the earthquake, and the failure of columns 4 and 5, were the causes of the collapse.

The Court reviewed the criticisms of the plaintiffs' experts:

(a) Computation of seismic loads did not consider lateral response.
(b) Splicing was in violation of AASHTO specifications.
(c) No provision for diagonal tension in anchorage.
(d) Footings and piles did not develop moment capacity of the piers.
(e) Expansion joint unrestrained.
(f) Inadequate seismic loads.

The Court then simplified the problem areas as basically two disputed aspects of the design:

(1) A correct analysis of the response spectrum of the

El Centro earthquake would have alerted designers to the intense lateral movements and that would have dictated a redesigned key.

(2) "Splicing" should not have been permitted at the column/foundation connections.

All other considerations were areas where engineers can (or at least in 1968 could) differ. The K factor, depth of footing, diagonal tension allowance, extra test concrete in some piers, skewing of piers, analysis of soil properties, larger restraints at the abutments, computation of periods for the entire structure of each half, various working stress levels, distribution of loads, were all matters in which there was reasonably different interpretation, or were factors discovered after the fact of the San Fernando earthquake.

Prior to the San Fernando earthquake the standard earthquake used was the El Centro earthquake of 1940. It was recorded with sufficient data to make it a model.

There was one group of experts which maintained that this bridge could have survived an El Centro intensity earthquake. Like so many engineering statements, it can only be proved empirically. The plaintiff maintained that it would not have survived El Centro, and in any event, an earthquake of greater intensity should have been anticipated in San Fernando.

THE COURT'S FINDINGS

The Court made the following findings. As to the hinge-key expansion joint: it was a standard feature that was common for all similar bridges. It had a standard 14" theoretical maximum expansion. In itself, it was widely accepted throughout the industry. Except for the unusual movements of this earthquake, there was no design imperfection in using this detail in this bridge.

As to the splice: the plans of the bridge made no reference to the word "splice." The joinder of the bars in the base of the columns and top of the piers was without direction. This was the area of failure, where the bars were pulled from the pile shaft. The Court found the practice at the time to be reasonable.

The considerations of the Court as to the earthquake factors were set forth in the findings. The greatest acceleration anticipated prior to the San Fernando earthquake was .5 G. The State of California in its bridge demands used seismic criteria patterned after the SEAOC Code. The equation EQ=KCD for the energy absorption of a structure (and using K=1.00 longitudinally and K=.67 horizontally) was not found to be an improper application. The Court found great assistance in the fact that the California bridge application as to earthquake factor was widely disseminated among structural engineers. No opposition to that treatment came to light prior to the reanalysis after the San Fernando earthquake.

EARTHQUAKE DESIGN REQUIREMENTS

The basis for rational seismic design was:

1. The structure remain elastic under influence of moderate earthquakes of frequent occurrence.

2. The structure should not collapse even under the most severe probable earthquake.

3. It was possible to meet these criteria based on an earthquake of the intensity and characteristics of the El Centro earthquake of 1940.

The California Department of Transportation bridge department met this design requirement as it was generally known in 1968. The 210/5 bridge was within these requirements as response spectrum analysis was then known. There was competent evidence that this bridge was designed to withstand acceleration of .5 G.

It was therefore the conclusion of the Court that the defendant had proven it was entitled to the defense of design immunity under Section 830.6.

The accident of February 9, 1971, was not caused by the faulty design of the bridge by then acceptable standards, and the design was approved in advance. The design was reasonable at the time under known laws of engineering, and probable anticipation of earthquake.

CITATIONS

1. California _Civil Code_, Section 1859

2. _Fay v. Pacific Improvement Co_. (1852) 93 Cal. 253, 261

3. _Ryan v. Rogers_ (1892) 96 Cal. 349, 353

4. _Ballentine's Law Dictionary_, Third Ed. 1969

5 _London Guarantee and Accident Co. v. Industrial Accident Commission_ (1927) 202 Cal. 239

6. California _Government Code_, Section 830.6

7. _Mikkleson v. State of California_ (1976) 59 Cal.App.3d 621

SEISMIC DESIGN FOR HIGHWAY BRIDGE FOUNDATIONS

Ignatius (Po) Lam[1] and Geoffrey R. Martin[2], M. ASCE

ABSTRACT

Experimental studies on bridge structures have shown that the lateral and rotational stiffness of pile groups may significantly influence the earthquake response. This paper focuses on the subject of the determination of stiffness characteristics of piles and pile groups, foundation types which are commonly encountered in bridge structures. Both battered and vertical pile groups are considered. Recommended methods for determining the axial and lateral stiffness of single piles are first presented, along with sensitivity studies to demonstrate the influence of liquefaction and strength degradation on lateral stiffness. Solutions are then presented showing how axial and lateral stiffness of individual piles may be combined to represent overall group pile stiffness to moment and lateral loading. Solutions for the particular pile group studied, indicate that significant bending moments may be induced in laterally loaded battered piles, and that seismically induced moment loading on a pile group may lead to permanent axial displacements.

INTRODUCTION

Highway bridges have been found to be highly susceptible to damage under earthquake loading. In terms of foundation performance, much of the easily recognizable damage has been associated with abutment failures or displacements during earthquakes or liquefaction of foundation soils (8, 11). In the absence of soil liquefaction and abutment displacement, damage to foundation elements such as column footings, piers and piles has been minimal. Nevertheless, the performance of these foundation elements can have a significant effect on the structural loading on the bridge during earthquakes. As a result, detailed consideration of the aseismic design of foundations is an important element of the overall seismic design process for highway bridges.

There has been a continual effort to upgrade the seismic design recommendations for highway bridges since the 1971 San Fernando earthquake (10). A major milestone in these efforts has been the completion of the Applied Technology Council study for the Federal Highway Administration (2). The seismic design guidelines resulting from this study have recently been adopted by AASHTO. While these guidelines are

---

[1] Senior Engineer, The Earth Technology Corporation, 3777 Long Beach Blvd., Long Beach, California 90807

[2] Vice President for Engineering, The Earth Technology Corporation, 3777 Long Beach Blvd., Long Beach, California 90807

fairly specific with respect to many of the structural aspects of the seismic design problem, they are less specific on many aspects of foundation design, where greater emphasis is placed on design philosophy and approach. This is in part due to 1) the complexities associated with performance of different foundation systems, 2) the wide variety of soil types encountered, and 3) the uncertainties associated with site specific soil properties.

Typical foundation systems include abutments, spread footings, battered and vertical pile groups, single column drilled piers and pile bent foundations (see Figure 1). Abutment types range from end diaphram to free standing abutments. Typical footing sizes range from 8 feet (2.4 m) to 30 feet (9.1 m) wide with thicknesses varying from 3 feet (0.9 m) to 6 feet (1.8 m). The commonly used pile types include timber, concrete and steel tubular piles with diameters ranging from 8 inches (0.2 m) to 24 inches (0.6 m) and pile length varying from 15 feet (4.5 m) to 30 feet (9.1 m). Typically piles are used in groups up to 6x6 piles. Center to center spacing in a pile group may range from 3 to 8 pile diameters. In the case of a battered pile system, batter angles of as much as 1 horizontal to 4 vertical are commonly used. Reinforced concrete drilled piers with diameters ranging from 4 to 8 feet (1.2 to 2.4 m) diameters with embedment lengths of 20 to 100 feet (6.1 to 30.5 m) are also commonly used for highway bridges.

The present design guidelines recognize two major aspects of foundation performance which impact seismic design: 1) the characterization and evaluation of the foundation system in terms of stiffness and damping properties to permit natural period evaluations or overall dynamic response of the bridge structure under earthquake loading, and 2) an assessment of the manner in which the earthquake loads are distributed to the individual member elements of the foundation systems and ultimately to the soil mass. Rather then giving a broad overview of the approaches used to evaluate these two aspects for all foundation types, we have chosen in this paper to focus attention on pile foundation stiffness characteristics, particularly pile groups which are commonly encountered in highway bridge structures. Recent experimental studies on highway bridges performed by Douglas, et al (6, 7) have clearly demonstrated the need to reliably determine both the lateral and rotational stiffness characteristics of pile groups in evaluating dynamic response.

For illustrative purposes, the paper focuses on the simple pile group system such as the one shown in Figure 2, where the foundation soils comprise a uniform sand. Current design procedure accounts for the foundation compliance by representing each foundation element by equivalent stiffness matrices (e.g. in Figure 3) for use in subsequent structural analysis. Such an equivalent stiffness matrix must be developed rationally. In the case of the pile group, the choice of the equivalent stiffness matrix should properly account for the following factors:

- The nonlinear nature of soil behavior, and variation of soil properties with depth.
- The configuration (spacing and batter angle) of the piles.
- The connection details between the piles and the pile cap.

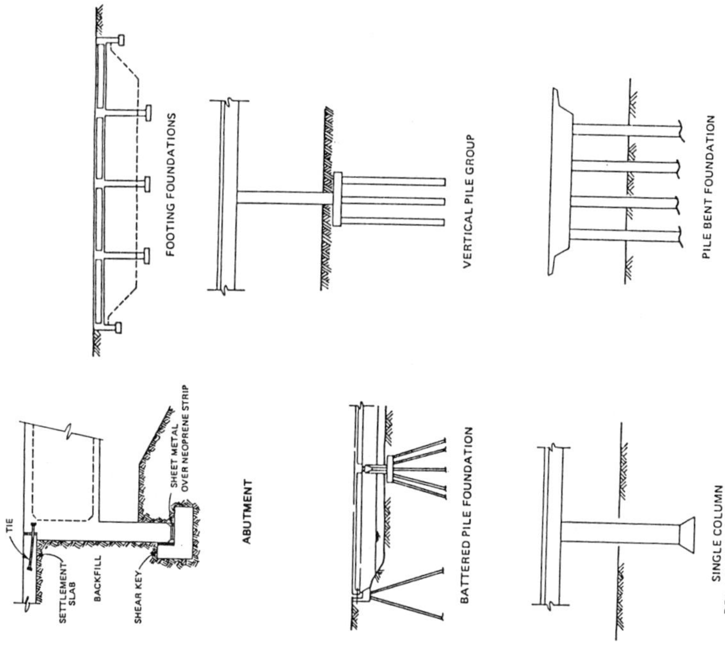

Fig 1. Representative Foundation Systems

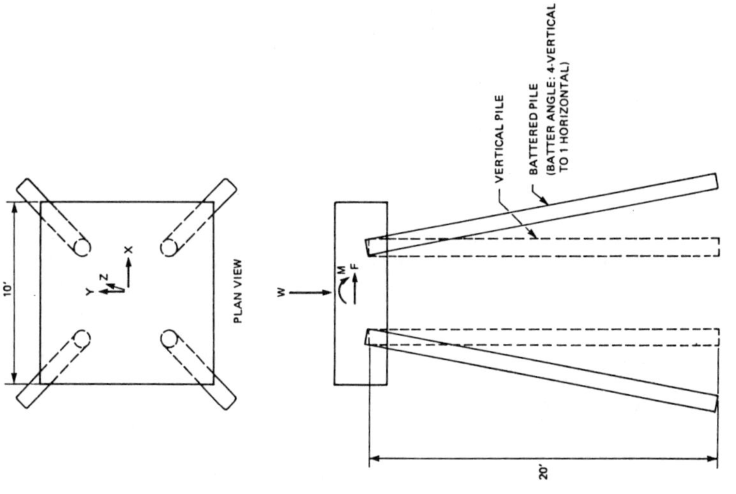

Fig 2. A Typical Pile Group

A series of analyses are presented in this paper to illustrate a recommended procedure to develop the equivalent foundation stiffness. Sensitivity studies are conducted to illustrate the significance of various factors (e.g. liquefaction, cyclic degradation, pile batter angles etc.) in affecting the pile group behavior. The subsequent paragraphs discuss the behavior of a single vertical pile under lateral and vertical loading and then progress to illustrate the behavior of a pile group.

SINGLE PILE BEHAVIOR

An appreciation of the behavior of a single pile is the most basic requirement in the design of piles and pile groups. A variety of methods are available for carrying out analyses of pile foundations. A realistic solution approach must consider the inhomogeneous layering and the nonlinear nature of soil behavior and yet be simple enough for practical applications. One of the more widely used practical approaches is the beam-column method (4, 15). This technique models the pile member as a series of beam column elements, with discrete springs to model the soil support (see Figure 4). With the above simplification, the inhomogeneous and nonlinear nature of soil behavior can readily be solved by a small computer, and very often by hand solution methods (12, 18).

Nonlinear discrete springs in the form of p-y (lateral support springs) and t-z (side friction springs) curves are usually used to simulate the lateral and axial soil supports (respectively) along the pile length. Past researchers have attempted to determine p-y and t-z curves using analytical methods (3) or alternately experimentally by conducting pile load tests (3, 13, 19, 20). Due to the complexity of soil behavior and the disturbance caused by pile driving, the empirical experimental approach for determining p-y and t-z curves remains the more reliable approach at the present time and has been adopted for example, by the American Petroleum Institute for offshore pile design (1). The following presents illustrative solutions for a single pile under various lateral, moment and axial loads. The parameters have been chosen to reflect the typical conditions encountered in highway bridge design.

Lateral Loading

The idealized pile and soil conditions used for the example problem are summarized in Figure 5. The family of p-y soil support curves at various depths along the pile was calculated based on the recommendations given by the American Petroleum Institute (1). Representative p-y curves at various depths are shown in Figure 6. To be realistic, the pile must be analyzed such that compatibility is maintained between the pile and the structure (e.g. at the connection point at the pile cap). Three different boundary conditions are normally assumed for load-deflection analyses of a pile-soil system. The pile top can be 1) fully free (zero moment), 2) partially restrained (some finite rotational constraint), or 3) fully fixed (zero slope). The fully fixed pile-head condition can rarely be achieved in practice. In reality some finite rotational constraint should be assumed even though the pile may be rigidly cast into the pile cap. All three types of connection (fully

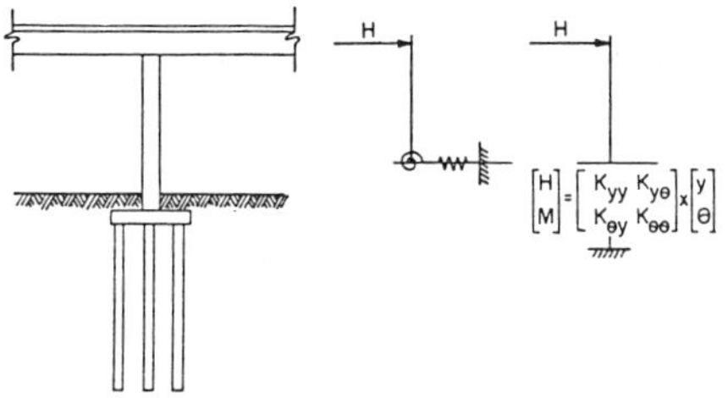

Fig 3. Foundation Stiffness

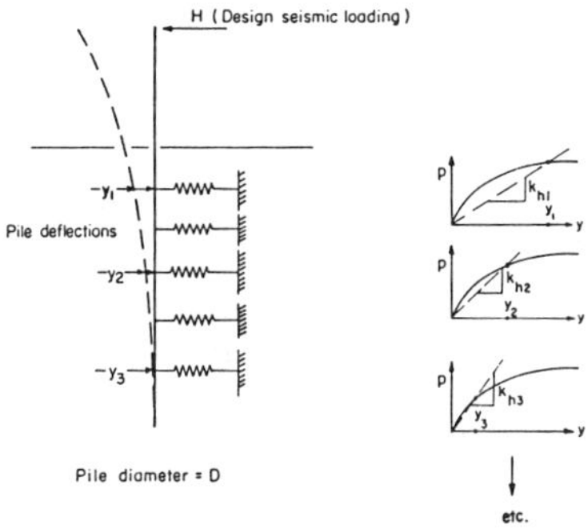

Fig 4. Soil Support Idealization

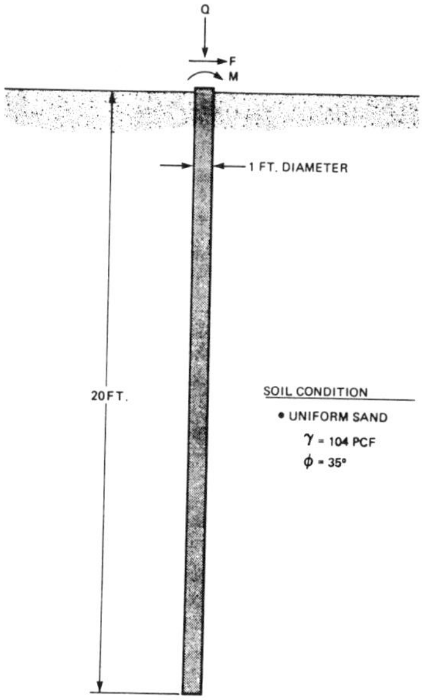

Fig 5. Pile Example for Analyses

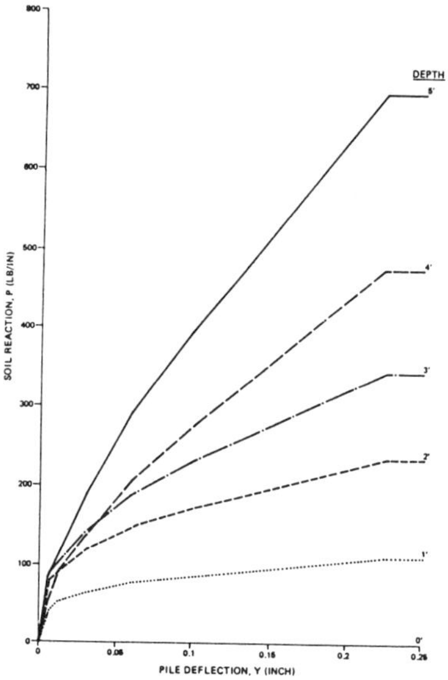

Fig 6. Family of p-y Curves
(1 in = 2.54 cm;
1 lb/in = 0.175 KN/m)

free, fully fixed and partially restrained) are analyzed in the following example and their implications compared and discussed.

Solutions were obtained for a range of lateral loads on the pile head using the BMCOL 76 Program (4,15). The resultant pile-head load versus pile-head deflection curves are plotted in Figure 7 for the three pile-head connections. In practice, it is commonly assumed that the free-head and the fixed-head pile assumptions serve as extreme bounds. The solutions support this assumption as the load-deformation curve for the partially restrained case falls between the free and the fixed head cases. However, this assumption is invalid for peak moment, as discussed below.

Figure 8 presents the cross-plot of maximum pile moment versus the corresponding pile head loads. This plot reveals that the pile-head load versus maximum moment curve for the partially restrained case falls below the range bounded by the free-head and the fixed-head curves. This anamoly is explained in Figure 9 which presents the moment distribution along the pile length for the three cases of pile-head conditions for a specific pile head load of 40 kips ($1.8x10^5$ N). A peak positive moment occurs at some depth below the pile top for a free-head pile, whereas, a peak negative moment occurs at the pile top for the fixed-head case. The partially restrained case gives a more balanced distribution of moment, where the peak negative moment at the pile top is roughly the same as the peak positive moment at depth. This balanced distribution results in a lower maximum moment as compared to either the free-head and the fixed head case.

Sensitivity to Soil Support

The p-y curves (See Figure 6) used in the above BMCOL solutions are based on the recommendations given by API for design of long offshore piles. Additional beam-column solutions are solved and presented below with the p-y curves modified to account for a number of soil-pile interaction aspects associated with earthquake considerations.

The partially-restrained pile-head condition is chosen for the sensitivity studies. The resultant beam-column solutions of pile-head load versus deflection curves are compared to the bench mark case previously described, and are presented in Figure 10. Some general comments of these comparisons are briefly discussed below.

P-Y Curve Shape. The API recommendations on the p-y curves place emphasis on the large deformation range (the ultimate resistance). Less emphasis is placed on the selection of the initial stiffness of the p-y curve as it has little influence on the pile solution for the large deformation range. Under earthquake loading, smaller pile deflections may occur, where the magnitude of the initial stiffness may warrant a more critical examination. To illustrate the effect of stiffening the initial soil response without changing the ultimate resistance, the benchmark p-y curve was modified such that the abscissa (the pile deflection) of the p-y curves were uniformly reduced to half the deflection values of the benchmark curve. As a result, the pile-head load-deflection curve could be expected to be stiffer than the benchmark

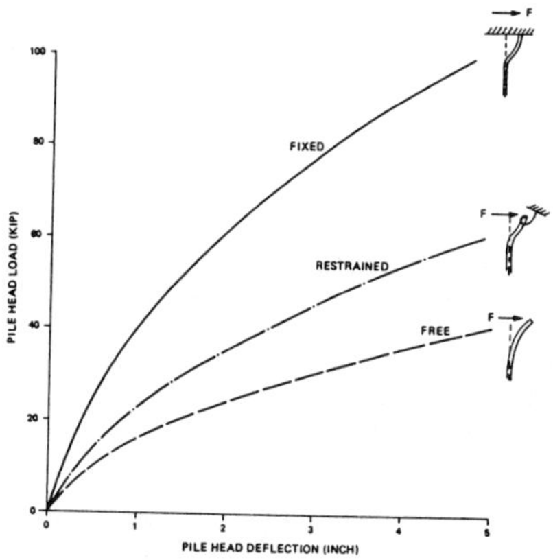

Fig 7 Load-Deflection Solution
(1 in=2.54 cm; 1 kip=4.45 KN)

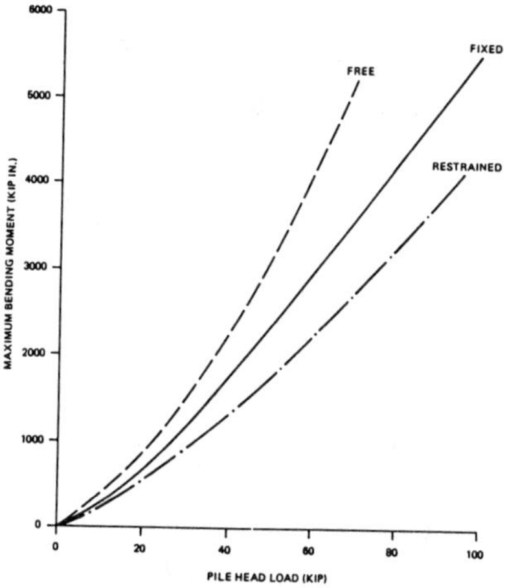

Fig 8 Moment-Load Solution
(1 kip=4.45 KN; 1 kip-in
1 kip-in = 0.11 KN-m)

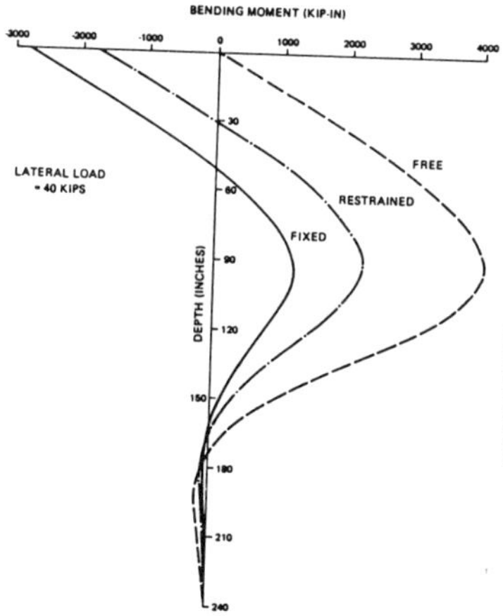

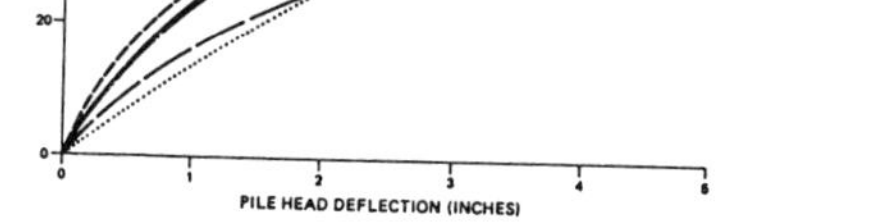

Fig 9 Moment Distribution
(1 in = 2.54 cm;
1 kip-in = 0.11 KN-m)

Fig 10 Sensitivity to p-y Curves
(1 in = 2.54 cm; 1 kip=4.45 KN)

case. However, the difference is seen to be very small, indicating that an accurate assessment of soil strength is more important then low strain stiffness in evaluating pile head stiffness.

Soil Gap Effects. During cyclic loading, the formation of a conical gap at the soil surface has been observed by a number of researchers (3, 13, 20). To simulate this gapping effect, one pile diameter (one foot depth) of the soil support (p-y curves) was eliminated at the soil surface. Below the one-foot depth, the same benchmark p-y curves were used. As shown in Figure 10, the resultant pile stiffness was only slightly reduced because of the gap effect.

Degraded p-y Curve. During cyclic and earthquake loading, the soil resistance may be progressively degraded by the effects of cyclic loading (13). The soil resistance on the p-y curves are reduced to half of the benchmark case, to arbitrarily account for potential cyclic degradation effects. The resultant pile-head load-deflection curve is shown in Figure 10. The resultant pile-head stiffness is significantly reduced especially at the small deflection range.

Liquefaction Effects. For loose sand deposits, the soil beneath the water table can potentially liquefy during an earthquake, resulting in a loss of soil resistance. To simulate a liquefaction condition, the p-y curves at zero to five feet depth were removed. However, the soil-support curves below five-foot depth were kept the same as the benchmark case. The resultant pile-head load-deflection solutions are presented in Figure 10. The pile-head stiffness is significantly reduced, especially at the small deflection range. However, the load-deformation curve becomes more linear, indicating that a higher proportion of the pile-head deflection is associated with the compliance of the cantilever beam over the liquefied zone.

The above sensitivity studies suggest that the cyclic degradation and liquefaction effects can potentially be more significant in affecting pile behavior and a greater emphasis should be placed on developing design guidelines to account for these aspects. Further studies illustrating the effects of liquefaction on pile response have been reported by Matlock et al (16) and Finn and Martin (9).

Axial Pile Behavior

In many cases, the lateral inertial forces of the highway bridge will introduce moment loadings on the foundation elements (e.g. pile groups). The rotational stiffness of a pile group depends strongly on the axial stiffness of individual piles. As a result, the axial pile behavior cannot be neglected in aseismic design considerations.

The axial load is transmitted by the pile member to the soil medium by way of skin friction along the pile shaft and by end bearing resistance at the pile tip. Thus, the pile-head stiffness (load-deformation) characteristics is affected by the following factors:

(1) the side-friction soil resistance versus pile displacement characteristics (side friction t-z curves);

(2) the end bearing soil resistance versus pile tip displacement characteristics (tip resistance t-z curve); and

(3) the elastic deformation of the pile member (the axial pile stiffness, AE, and the pile length L).

Simple computer solution methods (15) can readily be used to account for all the above factors. Simplified hand solution methods described below can also be developed based on an appreciation of the sensitivity of each of the above factors in affecting pile behavior.

A series of computer solutions are presented below for the previously described pile-soil system (see Figure 5). Simplified procedures for estimating axial pile stiffness are then developed from observations of the computer solutions.

The axial pile capacity curves for the sand profile are summarized in Figure 11 (based on the API recommendations (1)). At a pile penetration of 20 feet (6.1 m), the pile tip resistance during compressive loading is approximately 33 tons (2.9×10^5 N) and 65 percent of the total pile capacity. Under uplift conditions, the tip resistance is usually neglected and the uplift capacity is derived entirely from side friction.

To determine the pile-head stiffness, soil reaction versus pile displacement curves must be constructed for both side friction at each pile element (t-z curves) along the pile and for the tip resistance. The above family of curves were constructed for analysis. Representative curves are presented in Figure 12a. Curves were constructed such that the distribution of the ultimate side friction along the pile and ultimate tip resistance are consistent with the pile capacity curves summarized in Figure 11. The t-z curve shapes were chosen in accordance with the recommendations given by Vijayvergiya (22).

Computer Solution. The pile head load-displacement solution for a pile with an AE of 2×10^5 lb (8.9×10^5 N) was determined using the beam-column computer program (15). This pile stiffness reflects the most flexible pile type typically used in highway bridges. The resultant solution of the pile-head load-displacement curve under compressive and uplift axial load is shown as the dashed line in Figure 12b. The load-displacement characteristic is quite different between compression and uplift loading conditions for sand due to the large contribution of tip resistance.

Hand Solution. It is of interest to sum the ordinate (load value) of the accumulative side friction and the tip resistance curve. The resultant load deformation curve as shown in Figure 12b, will then represent a pile head load displacement curve when the axial pile stiffness (AE) is infinitely rigid. The difference between the rigid pile load displacement curve and the previously computed flexible pile solution represents the additional elastic compliance of the pile. Since the the flexible pile solution (dashed line) is computed using the most ductile pile likely to be encountered in highway bridge design, the bound between the rigid and the flexible pile load-displacement curves should bound the solutions for all the pile types used in highway bridges. Furthermore, from the above sensitivity study, a simplified hand solution method can readily be developed to solve for the pile head load-displacement characteristics of piles for highway bridge applications. The steps involved are outlined below:

(1) Construct the accumulative side-friction and the pile-tip resistance versus pile displacement curves, in accordance to the assumed distribution of side friction and tip resistance as well as the assumed soil-reaction displacement curve shapes.

(2) Construct a load-displacement curve by summing the load values of the accumulative side-friction and the tip-resistance curves at any displacement value resulting in a load-displacement curve that is compatible to the rigid pile case.

(3) Add an additional component of displacement to the above rigid load-displacement curve to account for the pile compliance. The amount of additional displacement at each load level (Q) can be approximated by the following equation.

$$\delta\text{(Pile Compliance)} = \frac{QL}{AE}$$

where

L is the pile length,
A and E are the cross sectional area, and the Young's modulus of the pile, respectively.

In most cases, except when the pile tip is embedded directly into bedrock, resulting in an extremely high tip resistance, the rigid pile load displacement curve (from Step 2 above) provides a good approximation of pile head stiffness. Most of the piles used in highway bridges are relatively short. Therefore, the component of displacement from the compliance of the pile (from Step 3) is relatively small.

PILE GROUP BEHAVIOR

The above discussion summarizes the fundamental behavior of a single pile. The behavior of a pile in a group may differ from a single pile due to stress transfer from neighboring piles through the soil medium. This is termed "group effect". Both the pile capacity and the pile stiffness can potentially be changed due to the group effect. Conventionally, group effects have been estimated from analytical solutions of a pile in an elastic continuum (17). These procedures in general overestimate the group effects. Experimental studies (14, 21) generally indicate that the group effect is insignificant for pile spacings (distance between piles) in excess of three pile diameters. At close spacing, adjustments on the p-y and t-z curves may be warranted to account for the pile group effects.

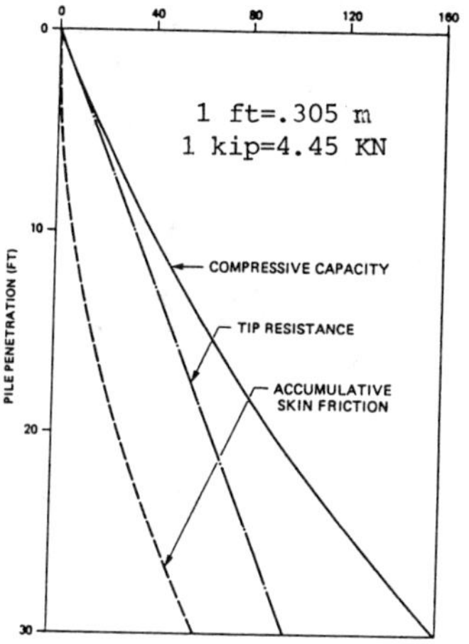

Fig 11 Axial Pile Capacity

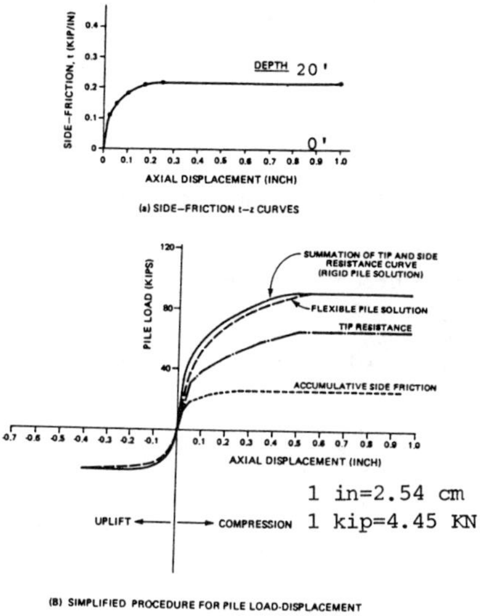

Fig 12 Axial Pile Solution

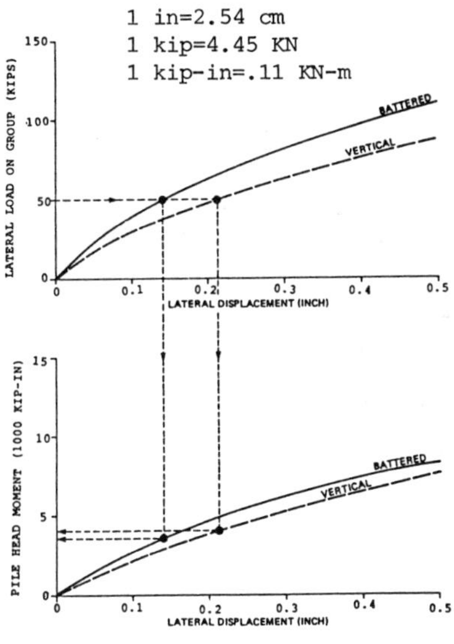

Fig 13 Group Solution
from Lateral Load

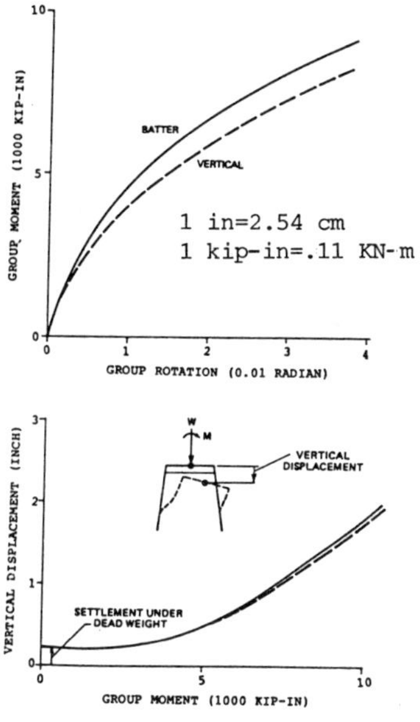

Fig 14 Group Solution from
Moment Loading

Pile Group Analyses

In addition to the "group effects" which influence the stiffness characteristics of an individual pile within a group, determination of the overall pile-group stiffness must consider the pile-cap interaction aspects including the configuration (spacing and batter angle) and the connection details between the piles and the pile cap. To illustrate the approach to account for all the above aspects in a pile group analysis, an example problem consisting of a 4-pile group in an uniform sand deposit is solved and described below. The configuration of the pile group is shown in Figure 2. Because pile spacing is greater than three pile diameters, group effects are not taken into account. Analyses for both four vertical piles and four battered piles are conducted and compared to evaluate the relative behavior of battered and vertical pile groups.

The pile-heads are assumed to be rigidly connected to the pile cap. The same sets of p-y and t-z curves used in the previous single pile analyses are used for this problem (Figures 6 and 12). The sets of loading considered in the analyses are 1) dead weight (W), 2) lateral force (F), and 3) moment load (M).

The above loads are applied to the pile group at the center of the pile cap. Two series of analyses are conducted: 1) monotonically increases of the lateral force (F) and 2) monotonically increases of the moment load. Both load types would arise from seismic loading of a bridge. In each series of analyses, a vertical load of 138 tons ($1.2x10^6$ N) is used to simulate the dead weight of the highway bridge. Under the above vertical load, 75 percent of the ultimate axial pile capacity is mobilized. This is higher than normal practice, but was used to illustrate potential impacts of moment loading on axial capacity.

Analytical Procedures. The above problem was solved using two computer programs STAR and PASS (5) which are run sequentially. The program STAR is a space-frame program used to model the pile cap or any space frame and to formulate the load vectors. The program PASS is used to model the four pile members and the nonlinear soil suports on each pile. Six degrees of freedom are used at each node in both STAR and PASS. In the solution process, the pile cap stiffness and load vectors are first condensed to the pile top using the STAR program resulting in a condensed superstructure stiffness matrix and load vector. Then, PASS is used to condense the stiffness of each pile individually to the pile-head. The condensed pile-head stiffness matrix is then added to the condensed superstructure stiffness matrix previously obtained from STAR. The resulting combined interface equations are then solved to obtain the interface (pile-head) solutions. Back-substitution along the piles yields the complete solution of each pile which can then be used to test against the nonlinear p-y and t-z curves. If convergence is not achieved, soil stiffnesses are then revised based on the revised displacement solutions along the pile length. The solution process is then iterated until convergence on the soil stiffness is achieved. Using the final converged interface displacements, back-substitution to the superstructure (in this case merely the pile cap) can then be performed.

Analytical Results. The two series of pile group solutions are presented in Figures 13 and 14 for the pile group under a lateral force and moment loading, respectively. The lateral stiffness for a battered pile group is expected to increase with an increase in 1) the battered angle, and 2) the axial stiffness of an individual pile. The axial stiffness characteristics of a single pile has been described previously. As shown in Figure 13, the lateral load versus pile group displacement for the battered pile group is 30 percent stiffer than the corresponding vertical pile group curve. The axial compliance of the pile in this example problem is largely due to the deformation of the soil. For some other situations, for example, if the pile tip is embedded into bedrock, the axial stiffness of the pile can be dramatically increased, and the lateral stiffness of a pile group can be significantly stiffened by the battered effect.

The bending moment on the pile-head induced by the lateral loading, differs only slightly among the four piles, and it is also plotted against the corresponding lateral displacement of the pile cap in Figure 13. In the design of a battered pile system, the shear force and moment at the pile cap is often assumed to be reacted primarily by the axial forces in the battered piles due to an assumed high axial pile stiffness. As a result the piles may not be designed for bending. This is clearly erroneous for non end bearing piles, where for a battered pile system, significant lateral load is taken out in bending due to lower axial stiffnesses. For a given lateral load (say 50 kips or $2.2\text{x}10^5$ N) the resulting bending moment for the battered pile system is indeed lower than the vertical pile group, however, a significant bending moment remains on the pile member. The above results suggest that bending moment in a battered pile system should not be neglected. Even if the pile-heads are designed for free head condition at the pile cap, significant bending moment can potentially occur at a deeper depth.

The pile group moment versus pile group vertical displacement solutions are shown in Figure 14 for the moment loading. Under the combined dead weight and the moment loading on the pile group, a significant amount of vertical displacement may occur at high moments as shown in Figure 14. The vertical displacement of the pile group is associated with the plunging failure of the compression pile, where the combined axial loading from dead weight and moment mobilizes the full axial capacity. A significant amount of the displacement could potentially be irrecoverable. Hence normal seismic design should allow for axial loading from seismically induced moment (that is, axial capacity not to be exceeded by dead plus seismic load) unless permanent displacements of the pile group can be tolerated.

SUMMARY AND CONCLUSIONS

Field experiments have clearly shown that the lateral and rotational stiffness characteristics of pile groups may significantly influence the earthquake response of a bridge structure. A recommended procedure to evaluate the lateral and rotational stiffness of a pile group taking into account the nonlinear nature of soil behavior, the spacing and batter angles of the piles in the group, and the connection detail between the piles and the pile cap is outlined in the paper. The sensitivity studies presented illustrate the importance of properly accounting for the three components of axial stiffness (side friction, end bearing, and pile stiffness), the potential effects of liquefaction and cyclic degradation on lateral stiffness, and the importance of properly combining lateral and axial pile stiffness to determine the overall stiffness of the pile group.

A wide range of sensitivity studies such as those described in the paper are being conducted for each foundation type commonly encountered in bridge structures. Based on these parametric studies, simplified procedures and design guidelines will be developed and documented for use in design practice. Although foundation behavior during earthquakes is recognized as being complex, results to date suggest that simplified solutions can be developed based on an appreciation of the sensitivity of dynamic response to foundation behavior.

ACKNOWLEDGEMENTS

The studies described in the paper form part of a broader program of study funded by the Federal Highway Administration, aimed at developing more specific guidelines for the seismic design of bridge foundations. These studies are being undertaken by a research team comprising engineers from The Earth Technology Corporation, Engineering Computer Corporation, and the University of Nevada, Reno.

REFERENCES

(1) American Petroleum Institute, API Recommended Practice for Planning, Designing, and Constructing Fixed Offshore Platforms. API, Washington, D.C., 13th Edition, 1982.

(2) Applied Technololgy Council, "Seismic Design Guidelines for Highway Bridges," ATC-6., 1981.

(3) Barton, Yvonne O., Finn, L. W. D., Parry, R. H. G., and Towhata, I., "Lateral Pile Response and p-y Curves from Centrifuge Tests," Proceedings, 1983 Annual Offshore Technology Conference, Texas, OTC 4502.

(4) Bogard, Dewaine, and Hudson Matlock, "A Computer Program for the Analysis of Beam-Columns Under Static Axial and Lateral Loads," Ninth Annual Offshore Technology Conference, Houston, TX, May 1977, Paper No. 2953.

(5) Bryant, Larry M., and Hudson Matlock, "Three Dimensional Analysis of Framed Structures with Nonlinear Pile Foundations," Ninth Annual Offshore Technology Conference, Houston, TX, May 1977, Paper No. 2955.

(6) Douglas, B. M., Norris, G. M., Saudi, M., Dodd, L. L., Richardson, J. A., and Reid, W. H., "Simple Bridge Models for Earthquakes and Test Data", Preprints submitted for publication in the American Society of Civil Engineering Journal of Structural Engineering.

(7) Douglas, B. M., and Richardson, J. A., "Maximum Amplitude Dynamic Tests of a Highway Bridge," Proceedings of the Eighth World Conference on Earthquake Engineering, July 21-28, 1984, San Francisco, California.

(8) Elms, D. G., and Martin, G. R., "Factors Involved in the Seismic Design of Bridge Abutments," ATC Workshop on the Research Needs of Seismic Problems Related to Bridges, Jan., 1979.

(9) Finn, W.D. Liam and Martin, Geoffrey R., "Offshore Pile Foundations in Sand under Earthquake Loading", Applied Ocean Research, Vol. 2, No. 2, 1980, pp. 81-84.

(10) Gates, J. H., "California's Seismic Design Criteria for Bridges," ASCE, Journal of the Structural Division, December 1976.

(11) Martin, G. R., "Seismic Design Considerations for Bridge Foundations and Site Liquefaction Potential," Proceedings, Workshop on Earthquake Resistance of Highway Bridges, Applied Technology Council, January 1979.

(12) Matlock, Hudson, and Lymon C. Reese, "Generalized Solutions for Laterally Loaded Piles," Journal of Soil Mechanics Division, ASCE, Vol. 86, No. SM5, October 1961, pp. 673-694; also in Transactions, ASCE, Paper No. 3370, Vol. 127, Part I, 1962, pp. 1120-1269.

(13) Matlock, Hudson, "Correlations for Design of Laterally Loaded Piles in Soft Clay," Proceedings, 2nd Annual Offshore Technology Conference, Houston, Texas, 1970, OTC 1204.

(14) Matlock, Hudson; Wayne, B. Ingram, Allen E. Kelley, and Dewaine Bogard, "Field Tests of the Lateral-Load Behavior of Pile Groups in Soft Clay," 1980 Offshore Technology Conference, Houston, Texas, 1980.

(15) Matlock, Hudson; Dewaine Bogard and Ignatius P. Lam, "BMCOL 76: A Computer Program for the Analysis of Beam-Columns Under Static Axial and Lateral Loading," In-House Documentation Report, Ertec, Inc., June 1981.

(16) Matlock, Hudson; Geoffrey R. Martin, Ignatius Lam and C.F. Tsai, "Soil-Pile Intgeraction in Liquefiable Cohesionless Soils During Earthquake Loading," Proceedings, International Conference on Recent Advances in Geotechnical Earthquake Engineering and Soil Dynamics, St. Louis, Missour, Vol. 2, April 1981.

(17) Poulos, H. G., "Group Factors for Pile-Deflection Estimations," Journal of the Geotechnial Engineering Division, ASCE, Vol. 105, No. GT12, Proc. Paper 15032, Dec. 1979, pp. 1489-1509.

(18) Reese, Lymon C., and Hudson Matlock, "Non-Dimensional Solutions for Laterally Loaded Piles with Soil Modulus Assumed Proportional to Depth," Proceedings, Eighth Texas Conference on Soil Mechanics and Foundation Engineering, Specialty Publication No. 29, Bureau of Engineering Research, The University of Texas, Austin, September 1956.

(19) Reese, Lymon C., Cox, W. R. and Koop, F. D., "Analysis of Laterally Loaded Piles in Sands, "Sixth Annual Offshore Technology Conference, Houston, Texas, May 1974, Paper No. 2080.

(20) Scott, R. F., Tsai, C. F., Steussy, D., and Ting, J. M., "Full-Scale Dynamic Lateral Pile Tests," Proceedings, 14th Annual Offshore Technology Conference, Houston, Texas, 1982, OTC 4203.

(21) Ting, John M. and Scott, R. F., "Static and Dynamic Lateral Pile Group Action," to be presented at the 1984 World Conference in Earthquake Engineering at San Francisco, preprint copy.

(22) Vijayvergiya, V. N., "Load-Movement Characteristics of Piles, "Paper presented in the Port 77 Conférence, Long Beach, California, March 1977.

SEISMIC CONSIDERATIONS FOR THE CENTURY FREEWAY

by James H. Gates* M. ASCE

ABSTRACT

The Century freeway (or Interstate 105), will be an 18 mile (29 km) long, six-lane facility with 2 central transit/HOV lanes, located in Los Angeles, California. This multimodal transportation facility with about $450,000,000 (US) in structures will extend in a general easterly direction from the Los Angeles International Airport and will require the construction of four major freeway interchanges and ten bus/rail stations. The freeway alignment crosses the Newport-Inglewood fault zone and seismic considerations are of primary importance in the design of the more than 200 structures in the project. The development of site-specific seismic response spectra for use in the design of the bridges is presented. The use of seismic resistant column and footing details based on the latest CALTRANS seismic criteria for bridges is discussed. The design considerations for large diameter pile shaft foundations are presented.

THE CENTURY FREEWAY

The Route 105 Freeway, (or Century Freeway), will be located in Los Angeles and extend from the Los Angeles International Airport eastward approximately 18 miles (29 km). The freeway will be six lanes wide with a 64 foot (20 m) median. There will be four major freeway to freeway interchanges constructed along the project (see Fig. 1.). A separated transit/HOV facility is planned to be located in the median with provision to convert to light rail. Ten bus stations with park and ride lots, designed to be convertible to rail stations will be constructed as part of the project. Local access to the freeway will be limited to ten locations along the length of the project.

Estimated project cost is about $1.5 billion (US), including approximately $450 million (US) for over 200 structures and $600 million (US) in right of way and relocation costs. The project is expected to be completed in late 1992.

---

*Senior Bridge Engineer, California Department of Transportation, Sacramento, CA.

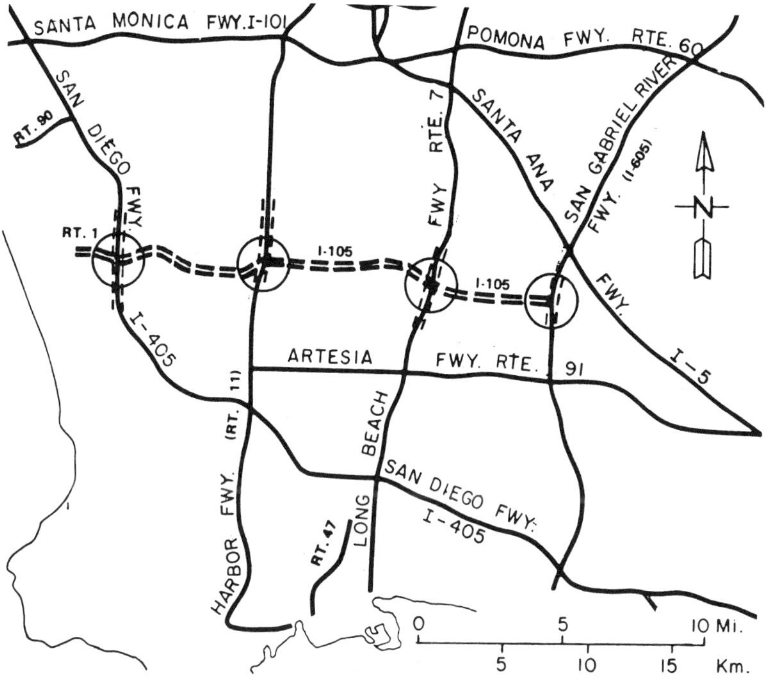

Fig. 1. Map of the Los Angeles Area Showing Location of the Century Freeway (I 105)

GEOLOGY

The Los Angeles basin is an ancient sea floor with alluvial and marine sediments estimated to be about 30,000 feet (9150 m) deep. These sediments overlie Mezozoic igneous and metamorphic bedrock (9).

There are up to 300 feet (92 m) of Recent alluvial deposits. These surficial deposits significantly affect the seismic response along the length of the project. The surficial deposits have been characterized into three regions (4):

1. Recent dune deposits at the westerly end of the project.
2. Upper-Pleistocene terrace deposits in the middle portion of the project.
3. Recent alluvial deposits at the easterly portion of the project.

The Recent dune deposits at the west end of the project consist of fine grained cohesionless material ranging from silt to medium sand. They are about 100 feet (31 m) thick and overlie consolidated upper-Pleistocene terrace deposits. No groundwater was encountered in this region.

The upper-Pleistocene terrace deposits in the middle portion of the project are highly variable. Generally they are composed of interbedded sands, gravels and silts. Some clay and clayey silt and peat were also found in this region. The only groundwater encountered in this region was some perched water in the vicinity of the 405/105 interchange.

The Recent alluvial deposits in the easterly portion of the project are made up of poorly consolidated sands, gravels, and silts. Thin scattered layers of clayey silt to clay were also encountered. The material varies in density from very loose to slightly compact at the surface with increasing density at depth and contains loose to compact layers to depths of about 80 feet (24 m). Groundwater depths vary from 20 to 60 feet (6 to 18 m) in this zone.

SEISMICITY

Eleven active faults were considered in the evaluation of the seismicity of the project area. One of these faults, the Newport-Inglewood, crosses the proposed freeway alignment in a NW-SE direction just west of the 105/11 (Harbor Freeway) Interchange.

A tabulation of the active faults considered showing the distances to the project area and assumed magnitudes and durations are shown in Table 1.

Table 1. -- Active Faults and Associated Magnitudes (4,10)

Fault	Distance to Proj. Limits Mi.	(km)	Maximum Historic Magnitude (Richter)	Maximum Credible Mag. (Richter)	Maximum Credible Duration (sec)
Cucamonga	24-40	(39-64)	---	6.6	18
Malibu-Santa Monica	8-18	(13-29)	5.2	7.5	30
Newport-Inglewood	3-12	(5-19)	6.3	6.5	24+
Palos Verdes	8-14	(13-23)	---	7.2	24+
Raymond Hill	12-14	(19-23)	---	7.5	30
San Andreas	37-44	(60-71)	8.3	8.3	40+
San Fernando	23	(37)	6.4	6.6	18
Santa Susanna	26	(42)	---	6.7	18+
Sierra Madre	18-24	(29-39)	---	6.6	18
Simi-Northridge	22	(35)	---	7.5	30
Whittier-Elsinore	8-14	(13-23)	---	7.6	30

Sensitivity studies were conducted to determine the effect of the distant faults on the motions at the project site. The computed motions from the San Andreas, Whittier-Elsinore and Palos Verdes faults were compared to the computed motions from the Newport-Inglewood fault. It was concluded that the motions from the nearby Newport-Inglewood fault represented a conservative envelope of the response spectra for the range of rock motions at the site. The procedures used to determine the computed motions are discussed under the topic 'SITE-SPECIFIC RESPONSE SPECTRA'.

FIELD INVESTIGATION

The field investigation for the ground response analysis was performed between April, 1981 and June, 1983. Sites for investigation were selected at each of the major interchanges plus two additional major structure locations (4).

A total of thirteen, 6 inch (152 mm) holes were drilled to depths of 200 to 350 feet (61 to 107 m). A small Concore drill rig was used to drill a pilot sampling hole to a depth of 100 feet (30 m). A Failing 1500 Holemaster was then used to drill the 6 inch (152 mm) hole and sample the material below the 100 foot (30 m) level.

Sampling was performed using a 1.4 inch (36 mm) split spoon sampler driven with a 140 pound (63.4 kg) hammer, (ASTM D 1586). A 2 inch (51 mm) modified California sampler fitted with brass tubes was also used to obtain samples for the determination of wet densities.

Shear wave velocity profiles as deep as 350 feet (107 m) were determined using the procedure developed and documented by Beeston and McEvilly in 1977 (1). This procedure determines the velocity of a surface generated shear wave which is propagated downward to a geophone located in the bore hole. Seismic waves were generated using hammer blows, delivered to the ends of a heavy plank loaded with the wheels of a vehicle. The waveforms were recorded on a seismograph at the surface. A series of records were obtained at various measured depths down the hole, permitting the computation of interval velocities. Shear waves are easily identified using this method by the presence of a clear 180 degree (3.14 radian) phase difference between waves generated by blows on the opposite ends of the plank. This method has been used at a number of sites in California and is capable of detecting a buried layer 5 to 10 feet (1.5 to 3 m) thick having a velocity differential of only 20%. A sample soil and shear velocity profile is shown in Fig. 2. Several soil and shear velocity profiles were averaged to obtain a series of composite profiles describing a series of zones along the length of project (See Table 2).

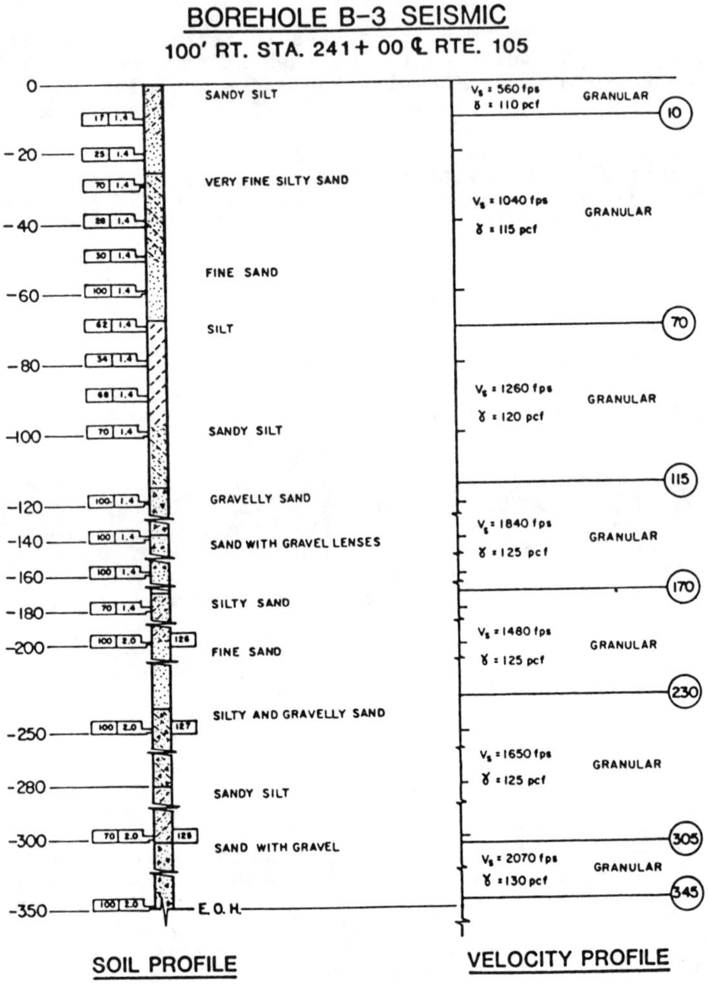

Fig. 2. Sample Soil and Shear Velocity Profile

SITE-SPECIFIC RESPONSE SPECTRA

The CALTRANS seismic criteria (3), permits the development of special design response spectra for specific sites. The spectrum is a 5% damped elastic acceleration spectrum which describes the maximum expected response at the surface of the site. This spectrum is referred to as an 'ARS' spectrum; where AR is the acceleration spectrum in rock-like material at the site and S is the soil amplification factor (8).

A total of six zones were selected to represent the seismic motions along the length of the project. (See Table 2) Boring logs for structures along the project length were reviewed to assure that computed ARS spectra would be representative of ground motions for all the sites within the zone.

Bedrock motions within the six zones were established by considering the closest distance from the causative fault to the zone and the maximum credible magnitude (See Table 1). The characteristics of the bedrock motions used for each zone were computed based on work by Seed et al. (17,18).

Table 2 -- Zones covered by ARS Spectra (4)

Curve No.	General Description	Coverage Mi.	(km)
1	Airport Viaduct	0.9	1.4
2	San Diego Frwy. (Rte 405) Interchange	3.9	6.3
3	Harbor Frwy. (Rte 11) Interchange	4.3	6.9
4	Alameda Viaduct	3.0	4.8
5	Long Beach Frwy. (Rte 7) Interchange	3.6	5.8
6	San Gabriel R. Frwy. (Rte 605) Interchange	2.5	4.0

Two earthquake time histories were utilized to describe the bedrock motions; a deconvoluted 1940 El Centro rock motion by Seed and an artificial rock motion generated by Romstad (15).

Surface motions were computed by propagating both of these rock motions to the surface of a one-dimensional soil column. The computer program, SHAKE (16) was used for this analysis.

The computation of the surface (or ARS) spectra was performed using a procedure similar to that used to develop the CALTRANS standard ARS curves (3,8). Recent investigators (2,11) have described this procedure as capable of obtaining a reasonable approximation of the motions at a site. The procedure is as follows:

1. Analyze a representative soil column for the bedrock motion using the SHAKE program (16).

2. Compute the 5% damped elastic response spectra for both the applied rock motion and the computed surface motion.

3. Compute the soil amplification curve (S curve) by dividing the surface spectrum by the rock spectrum.

4. Compute the surface ARS spectrum by multiplying the CALTRANS normalized rock spectra, R (8), by the peak rock acceleration at the site, A, and the computed soil amplification, S.

All soil columns were analyzed using a depth to rock-like material of 300 feet (92 m). This depth was selected after performing a series of sensitivity studies which evaluated the effects of depth variations to 600 feet (183 m) (4).

Smooth envelope spectra were prepared for each of the six zones. A comparison plot of the ARS curves developed for the project is shown in Fig. 3.

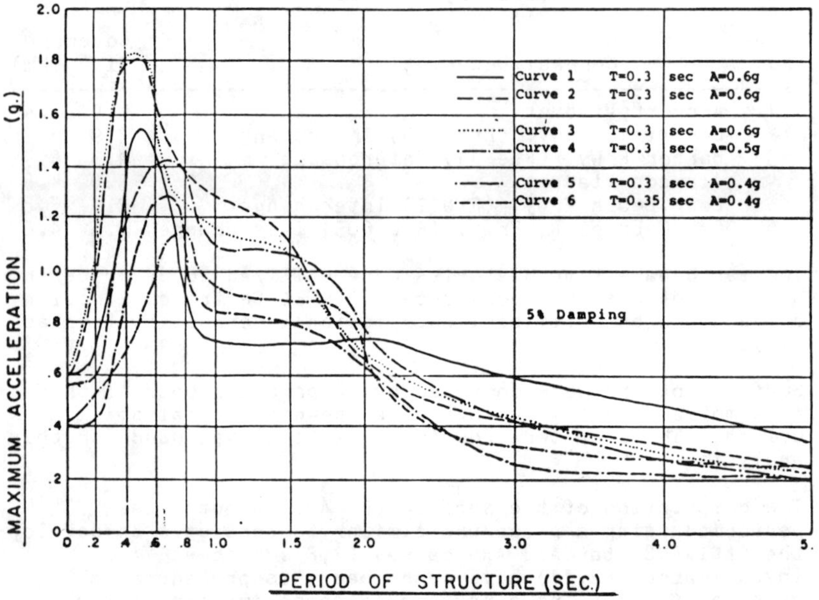

Fig. 3 Final ARS Curves

COLUMN AND FOOTING DESIGN

Recent criteria for the seismic resistant design of bridges (3,13) requires that any connections which are expected to undergo plastic deformation be able to develop the plastic moment and resultant shear as a limiting value. In California, this means that the plastic forces usually become the limiting design values to be used in the design of the connections. This load case when evaluated alone, tends to simplify the design of the connection.

In reality however, this requirement adds another level of complexity to the design process and the design engineer must now consider plastic loads in addition to evaluating loads at the factored and service level.

At CALTRANS the design of the column member is highly automated. An interactive computer program, YIELD, (5) is used to evaluate the column member for factored, service and plastic loads and determine vertical, shear and confinement reinforcement.

The current CALTRANS design specification (3), requires the extension of the confining spiral reinforcement into the adjacent cap or footing. The details for a typical single column/footing and column/cap connection are shown in Fig. 4.

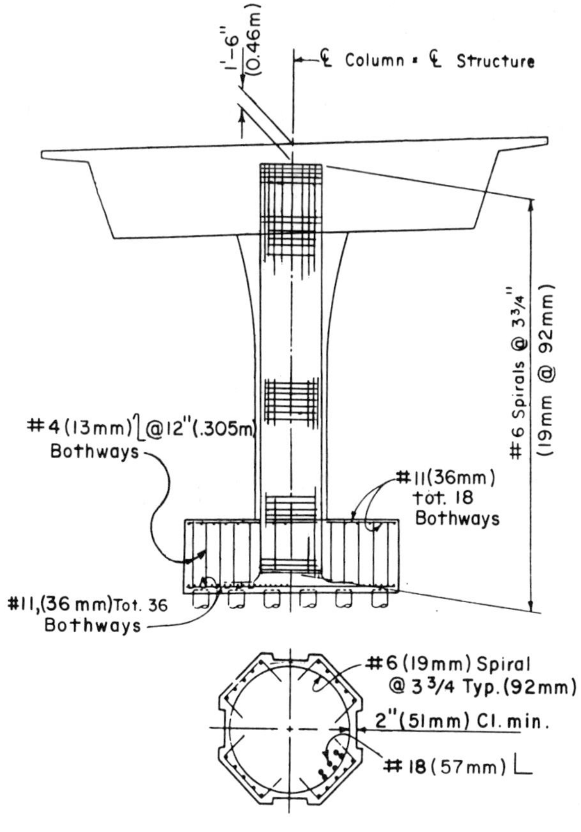

Fig. 4 -- Typical Single Column Bent on Footing

The Century Freeway column is generally octagonal in section with alternate recessed faces added for aesthetic emphasis (See Fig. 4). Shear and confinement reinforcement consists of continuous spirals (usually #6 or 19 mm bars). These spirals are spliced by welding or 80 diameter lap splices with 135 degree (2.36 radian), 10 inch (254 mm) hooks at ends. This will assure that the spiral will remain effective when the column is subjected to large excursions into the ductile range. Rectangular columns are reinforced for shear and confinement by using a series of interlocking spirals spaced not greater than 0.75 times the spiral diameter.

The design of the footing is also automated at CALTRANS using the program, FOOT, (6). Footings are evaluated for factored, service and plastic loads from the column. Plastic moments are applied at 15 degree (0.26 radian) intervals to consider the effect of plastic action in any direction.

Allowable soil stresses or pile loads vary depending on the class of loading (See Table 3).

Table 3 -- Allowable soil stresses or Pile Loads In Terms of Ultimate (3)

Foundation Material	Loading	Allowable (Times)	Ultimate
Soil	Service	0.33	Soil Stress
Soil	Factored	0.50	Soil Stress
Soil	Seismic & Plastic	1.00	Soil Stress
Pile	Service	0.50	Pile Load
Pile	Factored	0.75	Pile Load
Pile	Seismic & Plastic	1.00	Pile Load

Top footing reinforcement is designed for all imposed loads including uplift on piles and the weight of footing and overburden.

PILE SHAFT DESIGN

Drilled pile shaft foundations will be used at a number of locations on the Century Freeway project. The drilled shaft foundation is generally cheaper than other foundation types and permits the location of columns in tight locations with a minimum of disturbance to existing facilities. The use of this foundation is generally limited to areas where soil conditions permit economical excavation for the shaft and where ground water is not encountered. The presence of ground water does not prohibit the use of the drilled shaft, however the cost becomes considerably higher in this case.

It is more expensive to maintain the shaft walls underwater (usually thru the use of bentonite slurry) and underwater concrete placement is required. Vertical load tests by CALTRANS on carefully installed slurry seal piles have shown that the resulting pile is equal in strength to a normal dry constructed shaft.

Drilled shaft dimensions vary from 6 to 8 feet (1.8 to 2.4 m) in diameter and 30 to 60 feet (9 to 18 m) into the ground.

Construction of the foundation consists of drilling of the shaft excavation, installation of the reinforcement cage and placement of the concrete.

Column reinforcement is continuous but may be spliced by welding or approved connectors, however staggered lap splices are permitted for #11 (36 mm) bars or smaller in columns over 34 feet (10 m) high. Lap splices are not permitted near areas where plastic hinges are expected to form. Fig. 5 shows the reinforcement details for a typical 7 foot (2.1 m) diameter drilled shaft and column.

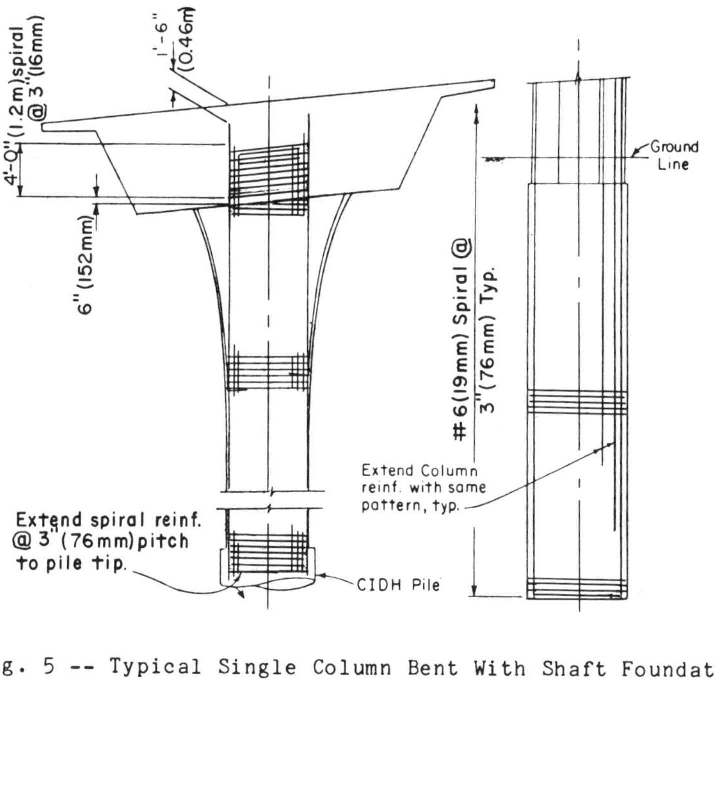

Fig. 5 -- Typical Single Column Bent With Shaft Foundation

The design problems involved with the use of the pile shaft are slightly different from those of ordinary pile and spread footing foundations. The pile shaft has a softer lateral stiffness and therefore requires more refined foundation data at an earlier stage in the design process. This softer lateral stiffness must also be considered in the design and analysis of the superstructure as well as the substructure components.

COLUMN/PILE SHAFT DESIGN PROCEDURE

The design procedure utilized for the Century Freeway pile shafts is as follows:

1. Determine an equivalent column length (Le) using the Kocsis procedure (12). The equivalent length (Le) of a column/shaft member is defined as that length of column plus pile shaft which when fixed at the bottom will produce the same deflections and rotations at the top for a given load as the actual column plus shaft surrounded by soil. The equivalent column has the same EI as the actual column and does not have any surrounding soil. (See Fig. 6)

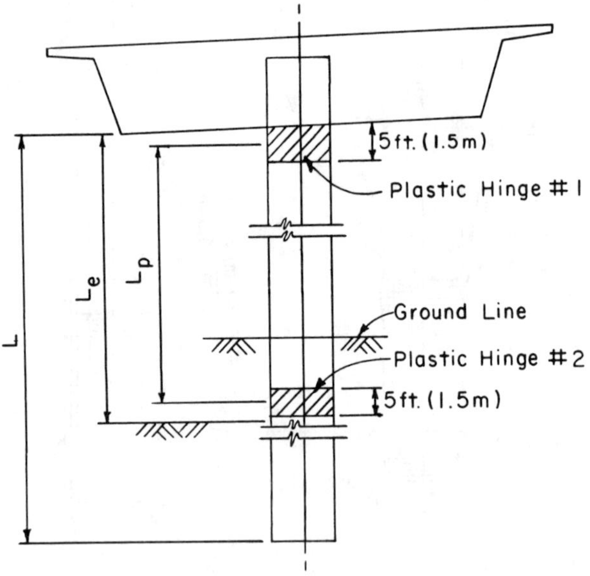

Fig. 6 --- Pile Shaft Equivalent Length and Plastic Hinge Locations

2. Perform a frame analysis using the equivalent column lengths (Le), to determine the maximum longitudinal forces and deflections (or rotations) at the top of the columns for a large service level load. Only one load is necessary for this analysis. The load is used to establish the level of the service load forces and deformations.

3. Obtain the soil data from a foundation investigation, including: soil classification, thickness of layers, soil density, modulus of horizontal subgrade reaction, minimum pile shaft length (based on maximum vertical loads) and internal friction angle (for sand), or undrained shear strength (for clay).

4. Analyze the pile shaft/column/soil configuration for the service level loads used in step 2. At CALTRANS this analysis is performed using the computer program COM624 developed by Reese (14,19). The total pile and column and surrounding soil are modeled in this analysis and the deflections and/or rotations at the top of the column are computed.

5. Check the equivalent length assumption, (Le), made in step 1, the rotation (or deflection) from step 4 is compared to the rotation (or deflection) computed in step 2. Generally the values should be within about 25% of each other to be assured of reasonable modeling. If necessary, adjust the equivalent length (Le) and re-run steps 2 and 4 until agreement is reached. Usually a reasonable amount of knowledge about the soils will permit the step 1 estimate of equivalent length, (Le), to be satisfactory for structural analysis purposes.

6. Design the vertical reinforcement required at the top of the column for all Load Groups (I thru VII). This design is performed using the YIELD program (5). Seismic analysis for Group VII is usually performed using modal analysis on a lumped mass model of the bridge. At CALTRANS this analysis uses the ICES STRUDL program distributed by the McDonnell Douglas Automation Company. The stiffness of the pile shaft foundation for STRUDL analysis is determined using the program FOSTIF (7) which computes the equivalent static and dynamic springs to model the pile shaft/soil configuration. Steps 1,2,4 and 5 can be eliminated for any analysis using stiffness analysis procedures by replacing the shaft/soil element with equivalent springs. Service level loads will require different springs than the large plastic or seismic loads.

7. Analyze the pile shaft for Group I thru VII loads using the COM624 program (14,19) described in step 4. The shears and moments along the length of the shaft are determined. The vertical reinforcement required for the maximum moment in the shaft can now be determined using the YIELD program (5). The plastic moment in the shaft is also determined based on the final reinforcement configuration. It will be assumed that a longitudinal plastic hinge will form at the top of the column and a second will form in the pile shaft (See Fig. 6)

8. Analyze the pile shaft for the plastic condition using the COM624 program (14,19). The plastic moment and the associated axial load and shear are applied at the top of the column. The program is then run interactively by incrementing the plastic shear until a second plastic hinge forms in the pile shaft. The shear reinforcement is then designed for this condition.

9. Perform a final check of the overall stability of the pile shaft using one of the following procedures:

   a. With the controlling shear, associated moment and axial load applied at the top of the column, maintain the pile shaft length and increase the shear until excessive deflection occurs. The pile shaft is considered stable when a substantial increase in shear does not result in excessive deflection. The amount of reserve deflection capacity is an indication of the factor of safety against overturning.

   b. With the controlling shear, associated moment and axial load applied at the top of the column, maintain the loading condition and decrease the length of the pile shaft incrementally until excessive deflection occurs. The pile shaft is considered stable when a substantial decrease in pile shaft length does not result in excessive deflection. The amount of reserve shaft length is an indication of the factor of safety against overturning.

APPENDIX - References

1. Beeston, H.E., and McEvilly, T.E., "Shear Wave Velocities From Down-Hole Measurements", Earthquake Engineering and Structural Mechanics, Vol 5, 1977, pp. 181-190.
2. Bell, J.M. and Hoffman, R.A., "Design Earthquake Motions Based on Geologic Evidence.", Proceedings of the ASCE Geotechnical Engineering Division Specialty Conference, Earthquake Engineering and Soil Dynamics, Pasadena, CA, June 19-21, 1978, Vol 1, pp 231-271.
3. "Bridge Design Specifications Manual", Standard Specifications for Highway Bridges Adopted by the American Association of State Highway and Transportation Officials, Twelfth Edition, 1977, With Interim Revisions Thru 1982, Washington, D.C., With Revisons by the California Department of Transportation, Office of Structures Design, Sacramento, CA, September, 1983.
4. "Century Freeway Ground Response Study", by the Office of the Transportation Laboratory, California Department of Transportation,Sacramento, CA, August, 1983.
5. "Column Design by Yield Surface 'YIELD'", Instructions to Users by the California Department of Transportation Office of Structures Design, Bridge Computer Manual, Chapter 5-3, July, 1983.
6. "Footing Analysis/Design Program 'FOOT'", Instructions to Users by the California Department of Transportation Office of Structures Design, Bridge Computer Manual, Chapter 8-2, July, 1983.
7. "Foundation Stiffness" by the California Department of Transportation Office of Structures Design, Bridge STRUDL Manual, Appendix E1, April, 1982.
8. Gates, James H., "California's Seismic Design Criteria for Bridges", Journal of the Structural Division, ASCE, Vol 102, No. ST12, December, 1976, pp. 2301-2313.
9. "Geotechnical Report for the Route 1/105 El Segundo-Norwalk Transportation Corridor", A Report by the District 7 Materials Section, California Department of Transportation, Los Angeles, CA April, 1974.
10. Greensfelder, R.W., "A map of Maximum Expected Bedrock Acceleration from Earthquakes in California.", California Division Mines and Geology, Map Sheet 23, Sacramento, CA, 1974.
11. Hays, Walter W., "Procedures for Estimating Ground Motions.", United States Department of the Interior, Geological Survey, Professional Paper 1114, Washington, DC, 1980.
12. Kocsis, Peter, "The Equivalent Length of a Pile or Caisson in Soil.", Civil Engineering Magazine, ASCE, Dec., 1976, pg. 63.

13. Mayes, Ronald L. and Sharpe, Roland L., "Seismic Design Guidelines for Highway Bridges.", Applied Technology Council, Palo Alto, CA and The Federal Highway Administration, Office of Research and Development, Washington, DC, Report No. FHWA/RD-81/081, October, 1981.
14. Reese, Lymon C. and Sullivan, W. Randall, "Documentation of Computer Program COM624", Geotechnical Engineering Center, Bureau of Engineering Research, The University of Texas, Austin, Texas, Aug., 1980.
15. Romstad, K.M., Bruce, J., and Hutchinson, J.R., "Site Dependent Earthquake Motions", Journal of the Geotechnical Engineering Division, ASCE, Vol 104, No. GT11, November, 1978, pp 1389-1400.
16. Schnabel, P.B., Lysmer, J., and Seed, H.B., "SHAKE: A Computer Program for Earthquake Response Analysis of Horizontally Layered Sites." Report EERC 72-12, Earthquake Engineering Research Center, University of California, Berkeley, CA, 1972.
17. Seed, H.B., and Schnabel, P.B., "Accelerations in Rock from Earthquakes in the Western United States." Report EERC 72-2, Earthquake Engineering Research Center, University of California, Berkeley, CA, 1972.
18. Seed, H.B., Idriss, I.M., and Kiefer, F.W., "Characteristics of Rock Motions During Earthquakes." Report EERC 68-5, Earthquake Engineering Research Center, University of California, Berkeley, CA, 1968.
19. "Shaft Pile Analysis", Instructions to Users by the California Department of Transportation Office of Structures Design, Bridge Computer Manual, Chapter 8-1, July, 1983.

SEISMIC DESIGN FOR SACRAMENTO LIGHT RAIL TRANSIT SYSTEM

by James E. Roberts[1], M. ASCE
and Robert E. Kershaw[2] M. ASCE

ABSTRACT

The Sacramento Light Rail Transit Project is an 18.5 mile system, primarily at-grade, which provides the trunk line service for an integrated Bus/LRT public transportation system for the metropolitan area. The system will operate eight 2 to 4 car trains on 15 minute headways to relieve bus congestion in the downtown core and reduce operating costs of the present transit system.

Seismic design for the structures utilized criteria developed for similar highway and railroad structures and typical industrial buildings. The major facilities include six new bridges, four bridge renovations, a large maintenance facility, passenger station platforms and shelters and the overhead electrical catenary support system.

This project is a typical "light rail" system in which the major facilities are minimized by utilizing existing right of way and at-grade crossings.

SYSTEM DESCRIPTION

The 18.5 mile Sacramento Light Rail Transit (LRT) Starter Line Project is being designed and constructed to provide the trunk line service for an integrated Bus/LRT public transportation system for the greater metropolitan area which has a population of 1.1 million and an annual growth rate of 2.5 percent. A single track main line will be built initially, with double track sections provided over 40 percent of the route to allow running meets between trains operating at 15 minute intervals (headways) in both directions. The design provides for future expansion to a fully double tracked system as ridership demand increases. This project is typical of European LRT design in which major structures are minimized by utilizing existing rights-of-way and at-grade crossings. The integrated

---

[1] Project Director, Sacramento Transit Development Agency (STDA) Sacramento, California.

[2] Deputy Project Director, Manager, Structures & Facilities (STDA)

Bus/LRT system will operate eight 2 to 4 car trains on 15 minute headways on the trunk line to relieve express bus congestion in the downtown core and reduce operating costs of the present all bus public transit system. The bus element will operate some 230 buses to provide feeder service to six "timed transfer" stations, thus maintaining the flexibility of buses to adjust routes as population patterns shift. The combined system will be operated at the same annual operating budget as the present all bus transit system because of LRT operating economies from lower operating costs and higher passenger, driver ratios. Total capital project costs for this LRT system are approximately $8 million per mile for double track compared to $50 to $100 million and higher for double tracked "heavy rail" systems which are completely grade separated. The tradeoff for this lower cost is reduced "level of service" but in most western U.S. cities, population density does not warrant the more expensive "heavy rail" type public transit systems.

The Sacramento LRT system will follow the two major transportation corridors along Interstate 80 from the Northeast and U.S. 50 from the East to feed commuters into the downtown core. See System Map, Figure 1.

The Starter Line Project will begin at Watt Avenue and I-80 in the Northeast Corridor. It will follow the abandoned I-80 Bypass freeway right-of-way (R-O-W), the abandoned Sacramento Northern Swanston Branch R-O-W along Arden Way, Del Paso Boulevard, the Route 160 bridge across the American River, 12th Street, K Street, 7th Street (southbound) and 8th Street (northbound), O Street, 12th Street, Union Pacific R-O-W adjacent to the alley between Q and R Streets, R Street, and the Southern Pacific Placerville Branch R-O-W in the Folsom Corridor to Butterfield Way. Future expansion to the Northeast, East and South are in the current planning program. Acquisition of critical rights-of-way on the initial starter line provide capability for easy expansion to a fully double tracked system.

A total of 27 passenger stations will be provided, six (6) to include bus transfer facilities, and seven (7) to include automobile park-and-ride lots. Outlying stations will have bicycle parking facilities where appropriate. A yard and shop complex will be located in the I-80 Bypass R-O-W near Academy Way between El Camino and Marconi Avenues.

Vehicles and other critical components and materials are on order or already delivered. Construction is underway and the first leg (the Northeast Corridor), will be completed by September, 1985. Initial revenue service is scheduled for November, 1985.

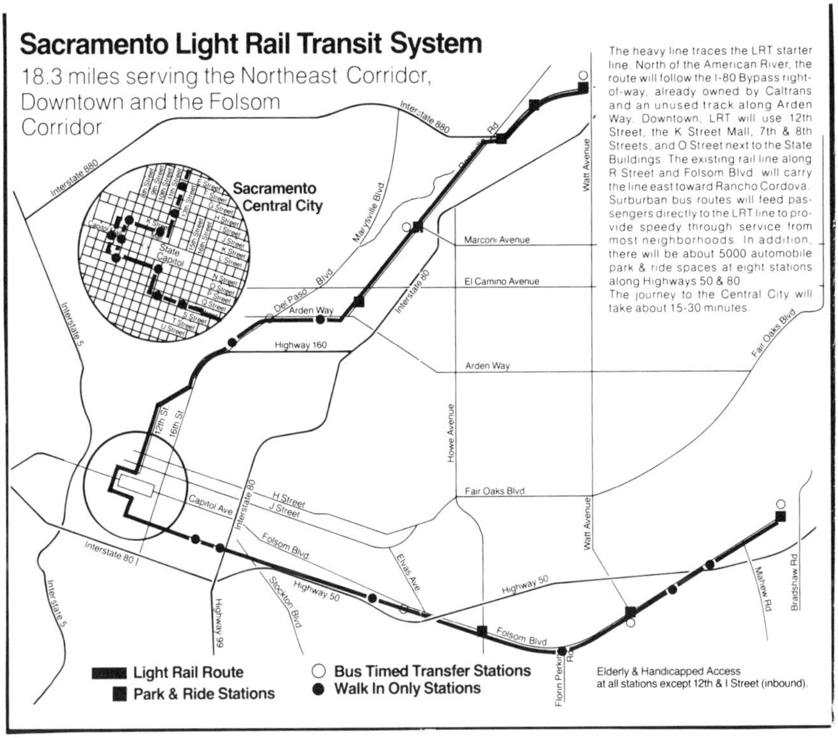

FIGURE 1 SYSTEM MAP

SEISMIC DESIGN CRITERIA

Because of the "at grade" nature of a typical LRT system major structures are held to a minimum. The major structures on the Sacramento System include three (3) highway/-LRT grade separations, one (1) LRT stream crossing bridge, two (2) major LRT/Freight Railroad Separations, four (4) existing bridge renovations to provide LRT access on previously constructed highway bridges, one (1) large LRT vehicle maintenance building, several passenger station platforms and shelters and the overhead D.C. power catenary support system (OCS). Seismic design for all these structures utilized criteria developed for similar highway and railroad bridges and typical industrial buildings and structures.

Bridge structures are designed to resist earthquake motions by considering the relationship of the site to active faults, the seismic response of soils at the site, and the dynamic response characteristics of the total structure.

Sacramento is located in the Central Valley on a deep alluvial deposit where the maximum credible rock Acceleration Coefficient (A) is 0.20 as shown on the map in Figure 2.

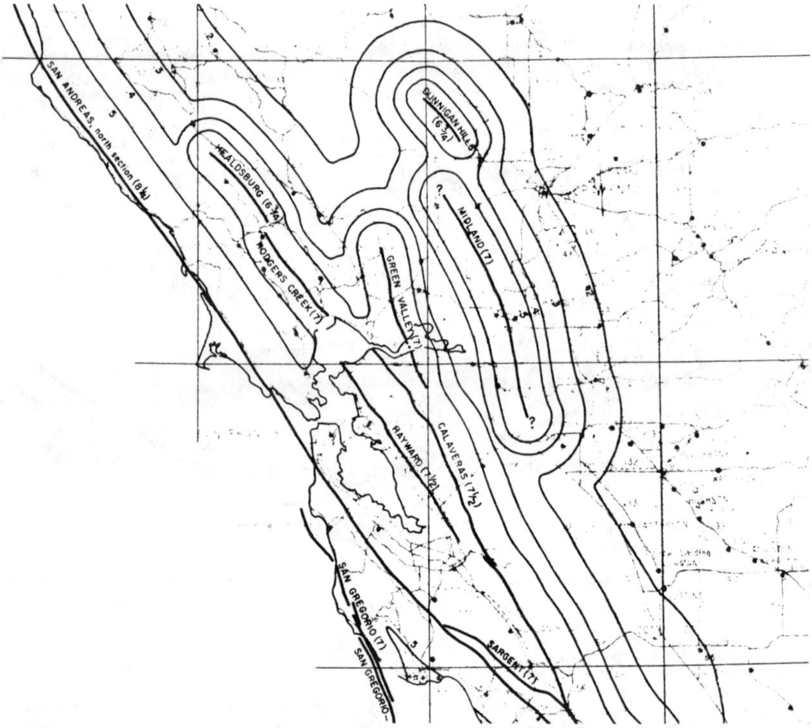

FIGURE 2 MAXIMUM CREDIBLE ROCK ACCELERATION

This coefficient is used for bridge design. For buildings and the OCS supports the seismic design factors are derived from the Uniform Building Code (UBC) seismic zone map shown in Figure 3.

During the project Preliminary Engineering phase a Design Criteria manual was developed for use by staff engineers and consultants. Excerpts of the structural section of the Criteria are included as Figures 4 and 5. The LRV design vehicle produces a live load on structures very similar to an AASHTO HS20 truck. Since live load is not included in the combination of loadings for seismic design, this fact is only significant because the structural design results in a bridge which is proportioned almost identical to a highway structure with the same geometric characteristics.

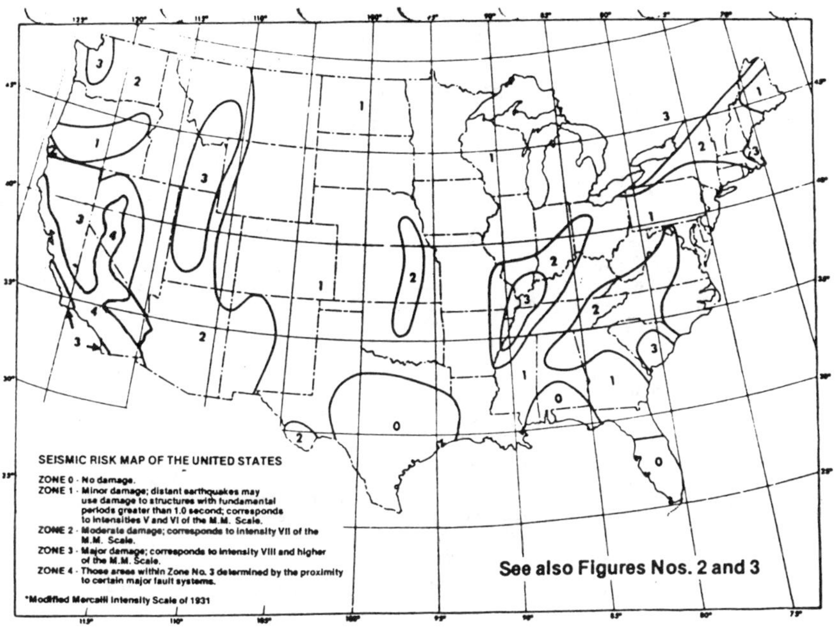

FIGURE 3 SEISMIC ZONE MAP OF THE UNITED STATES

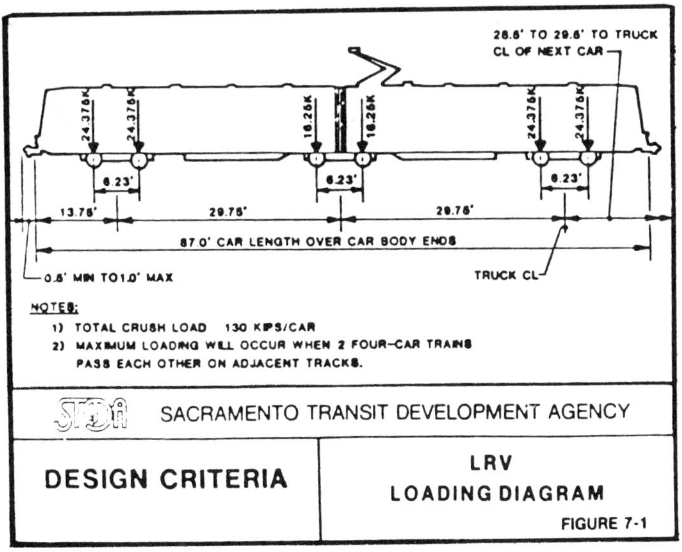

FIGURE 4 LRV DESIGN VEHICLE

7.0 STRUCTURAL

These criteria shall govern the design of structures constructed as part of Sacramento's Light Rail Transit (LRT) project.

Design loads for bridges, tunnels, culverts, and retaining walls shall be as specified in the AASHTO Standard Specifications for Highway Bridges, supplemented by Caltrans Bridge Design Specifications (BDS) and the following provisions for rail transit facilities. In case of conflict among these sources, provisions of BDS govern over AASHTO, and the rail transit provisions govern over those of both BDS and AASHTO.

7.1 DESIGN CODES, MANUALS, AND SPECIFICATIONS

Unless otherwise specified herein, the structural design shall be guided by the following codes, manuals, or specifications as appropriate:

7.1.1 Buildings

Uniform Building Code, 1982 Edition and local building codes as appropriate.

7.1.2 Railroads Bridges

Design requirements of the applicable railroad. In the absence of such requirements, the "Manual of Recommended Practice of the American Railway Engineering Association", hereinafter referred to as the AREA Manual.

7.1.3 Highway Bridges

The following codes and specifications shall be adhered to:

American Association of State Highway and Transportation Officials (AASHTO) Standard Specifications for Highway Bridges, Twelfth Edition, with all Interim Supplements and the California Department of Transportation Bridge Design Specifications (BDS) modifications of the AASHTO specifications.

7.1.4 Light Rail Structures

Caltrans Rail Transit Criteria Report No. UMTA/CA/PD-78/1, as modified by the design vehicle shown on Fig. 7-1 and Sections 7.2, 7.3 & 7.4 herein after.

12/29/82

FIGURE 5 DESIGN CRITERIA

The Caltrans Bridge Design Specifications (BDS) used by STDA in our Design Criteria provide a method for determining response spectrum loading of the ground acceleration on a lumped sum mass space frame model of the structure by using the A•R•S curves in Figure 6. In the derivation of these curves, A is the maximum expected acceleration at bedrock, R is the normalized rock response and S is the soil amplification spectral ratio. A•R•S is the 5 percent damped elastic acceleration response spectrum.

The curve for Sacramento is for alluvium greater than 150 feet in depth above the base rock. Entering the curves with peak base rock acceleration for the site, A (0.20)

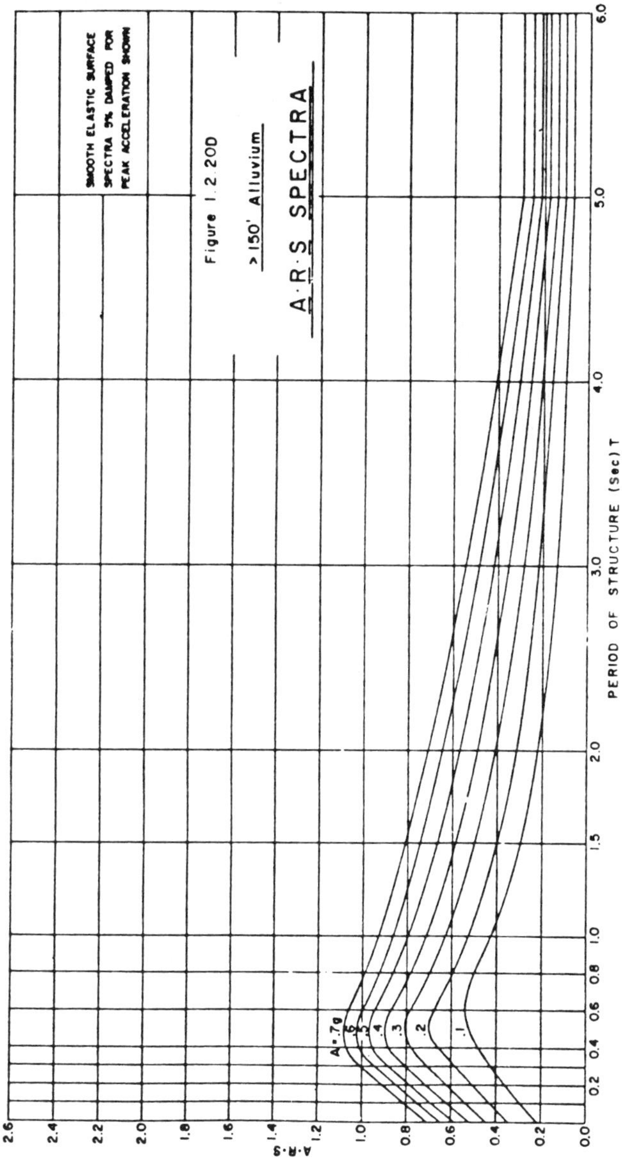

FIGURE 6 A•R•S SPECTRA

and the fundamental period of the structure, T (0.07), (approximate value for most of the larger Sacramento LRT structures) the designer determines peak acceleration to be applied to the structure. For this example that peak acceleration factor A•R•S is 0.65g.

The combined response coefficient (C) is determined by dividing the resulting elastic acceleration response spectrum by a Ductility and Risk Assessment factor (Z) from Figure 7. For a typical single column bent single track bridge the Z factor is 6 and the acceleration factor of 0.65g is reduced to C=0.11g for a structure where T=0.7 seconds.

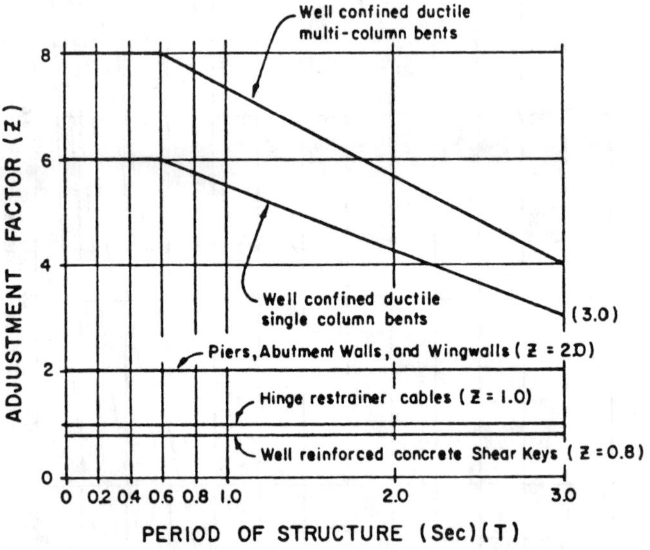

ADJUSTMENT FOR DUCTILITY AND RISK ASSESSMENT

FIGURE 7

Seismic design forces (EQ) are determined by applying the combined response coefficient, C, to the dead load (W) of the bridge. In order to account for directional uncertainty of earthquake motions, the seismic forces resulting from longitudinal and transverse loading conditions are combined for two loading cases:

EQ load Case 1 - Combine longitudinal seismic design forces with 30 percent of the transverse forces.

EQ load Case 2 - Combine transverse seismic design forces with 30 percent of the longitudinal forces.

For bridges with well balanced spans and supporting bents or piers of approximately equal stiffness, EQ may be applied as a static force at the vertical center of gravity (CG) of the total structure in any horizontal direction.

For bridges with significantly irregular configuration or support stiffness, EQ may be determined directly as individual member forces by dynamic analysis. This response spectrum method utilizes a modal analysis based on the application of a response spectrum of ground acceleration to a lumped sum mass space frame model of the structure.

HIGHWAY GRADE SEPARATIONS

These three structures are four lane, multi-span cast-in-place prestressed concrete (CIP/PS) box girders to carry local city streets over the LRT, Southern Pacific Mainline and local frontage roads. All design for these structures including seismic design, was implemented in accordance with established AASHTO and California Department of Transportation Bridge Design Specifications (BDS). Another paper presented in this symposium covers highway bridge design in detail.

LRT STREAM CROSSING BRIDGE

This bridge is a 130 foot long CIP/PS concrete box girder double track structure across a small creek. A simple span was selected to eliminate working in the channel for environmental reasons. Since the vehicle live load closely approximates the conventional highway type loading, the design was similar to highway structures. Because it is a single span a detailed seismic analysis was not required. Other loadings are as noted or specified in the AASHTO Standard Specifications for Highway Bridges and supplemented by the Caltrans Bridge Design Specifications.

LIGHT RAIL VEHICLE/FREIGHT RAILROAD SEPARATIONS

Two grade separation structures are required to eliminate conflicts with operating railroads. Each of these structures are approximately 1400 feet long, single column, multi-span, cast-in-place, prestressed concrete (CIP/PS) box girders providing a single LRT track with direct fixation to the deck. See Figures 8 and 9.

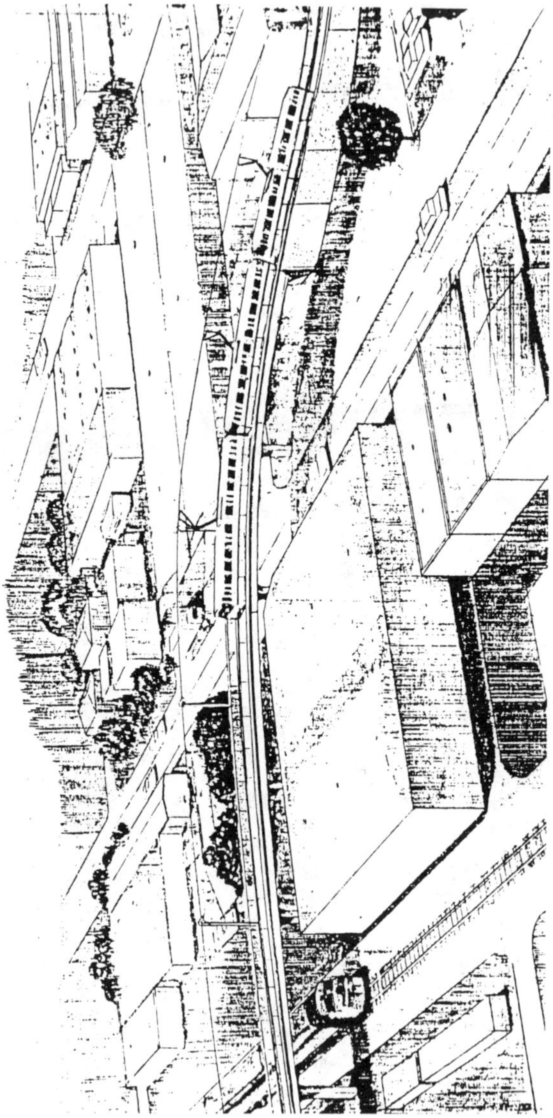

FIGURE 8 LRT/FREIGHT SEPARATION

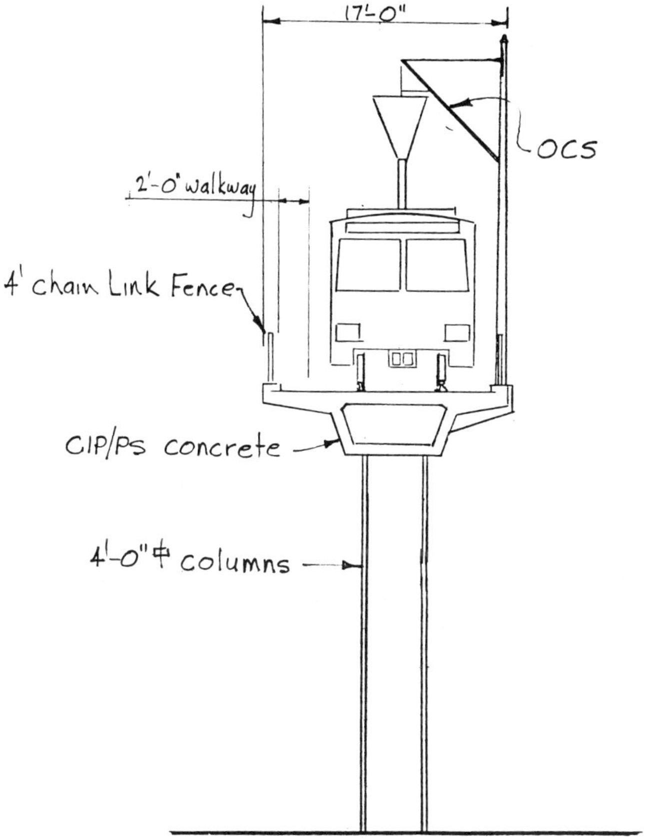

FIGURE 9 TYPICAL SECTION

Present California bridge design practice requires a rigid connection between superstructure and top of columns for maximum stability. That same practice eliminates hinges at the bottoms of single column bents. For longitudinal stability a minimum of two columns per frame are strongly recommended.

Hinges are kept to a minimum and hinge seat width is 24 inches. Cable restrainers are utilized to keep hinges together during seismic activity. These restrainers are designed for the full seismic design peak acceleration factor A•R•S=0.65g because the Ƶ factor from Figure 7 is 1.0, therefore providing no reduction. Shear keys in the hinges and abutments are designed for an A•R•S=0.80g because the Ƶ factor is only 0.8.

Analysis of the structural frame and review of site conditions indicate a structure period of 0.34 seconds longitudinal and 0.70 seconds transverse. The alluvium depth is greater than 150 feet. The maximum credible horizontal acceleration of the base rock is A=0.29. Using the appropriate A•R•S table (Figure 6) the peak acceleration to be applied on the structure is A•R•S=0.65g transverse and 0.62g longitudinal. For deflection investigation these factors are not reduced for ductility and the forces calculated from contributing dead load and A•R•S peak acceleration are applied statically at the center of gravity of the superstructure mass. For structural calculations the A•R•S factors are reduced by the ductility factor of 6 for transverse and 8 for longitudinal forces. The resulting adjusted acceleration factors (combined response coefficients) are C=0.11g for transverse design and C=0.08g for longitudinal design. These coefficients are applied to the contributing dead load to obtain EQ for both transverse and longitudinal directions. The two EQ load cases are then analyzed so the structure may be designed to resist the combined loads. Several iterations utilizing varying stiffnesses may be required to arrive at the final design.

Hold down devices are provided at all supports and hinges where the vertical seismic force exceeds 50 percent of the dead load reaction.

HIGHWAY BRIDGE MODIFICATIONS

Three major highway structures are being modified to accommodate light rail. One structure is a multi-span concrete highway bridge that was to have been used for a planned freeway extension. The 60 foot wide structure will now be used for a single LRT track and a two-lane parking lot access roadway. This structure was designed in 1968 and incorporated the state-of-the-art earthquake design criteria in effect at that time.

A recent dynamic analysis of the structure indicates no need for earthquake restrainers at hinges and no need to retrofit as part of the bridge earthquake upgrade program for substandard structures. This structure location has a maximum bedrock acceleration of A=0.2g and alluvium depth of less than 60 feet. This site also has a very low potential for liquefaction.

The other two structures are early to mid 1900 multi-span concrete structures crossing a flood plain and the American River. These structures have since been modified or widened to accommodate increasing vehicular traffic. Both will continue to carry three vehicular traffic lanes as well as a single track light rail line with direct rail fixation to the deck.

A recent dynamic analysis was performed on both structures and it was determined that no earthquake restrainers or other structural upgrading was necessary as all current criteria for earthquake resistance was met.

MAINTENANCE BUILDING

The building is located in Seismic Zone 3, and was designed using the 1979 UBC. The 50,000 sq. ft. structure is divided by major work functions: the main service area, administrative offices, heavy component repair shop, storage and repair shops, and paint and body shop.

Because the structure is basically a single story open bay, the seismic design utilized the "static force method."

The lateral loads in the administration section and storage and repair shops are resisted by reinforced shear walls. The shear walls are either tilt-up concrete panels or masonry. A Horizontal Force Factor "K" of 1.33 was used to determine the base shear. These sections of the building have relatively minor horizontal deflections due to seismic loads.

The main service area is a structural steel rigid frame type of construction with a clear span of approximately 88 feet. The walls for the building are tilt-up reinforced concrete panels. The lateral loads parallel to the rigid frames are carried by the rigid frames and the lateral loads perpendicular to the rigid frames are resisted by either tilt-up panels or steel portal frames. To determine the base shear for this portion of the building, a Horizontal Force Factor "K" of 1.00 was used for the rigid and portal frames and 1.33 for the tilt-up walls. Since the horizontal deflection of the steel frames due to seismic loads is larger than the shear wall deflection, the two types of building construction were separated by a space of two inches. This space will prevent damage to the building due to the different magnitude of horizontal deflections. A joint had to be designed to take this movement.

The building has precast concrete fascia panels attached to the columns and the tilt-up panels for architectural purposes. The cantilevered panels were designed using a horizontal force factor "Cp"=0.8 and the remaining panels with Cp=0.3.

PLATFORM SHELTERS

The transit shelters are composed of cold formed structural steel tube framing system with concrete columns. The structures are open and relatively small, light weight. Because of this, wind forces on the exposed structure framework controls over seismic design and determine the final structural configuration. The canopies are influenced by seismic design at the connections and columns. The code requires a positive connection to the roof/canopy at the top of columns. Column and foundation design is controlled by wind forces.

OVERHEAD CATENARY SYSTEM (OCS)

The OCS is generally supported by single cantilevered steel poles spaced about 100-150 feet along the route. These poles and foundations are designed to resist the catenary wire forces due primarily to temperature, wind and seismic loads.

A soils report was prepared based on information obtained from rotary borings taken along the route. The soil parameters obtained are used in the "Designs Employing Lateral Bearing" Section of the Uniform Building Code. This "flagpole" analysis determines the pole anchorage, foundation requirements and pole structural section.

APPENDIX - References

1. "Bridge Design Specifications Manual," Standard Specifications for Highway Bridges Adopted by the American Association of State Highway and Transportation Officials, Twelfth Edition, 1977, With Interim Revisions Through 1982, Washington, D.C., With Revisions by the California Department of Transportation, Office of Structures Design, Sacramento, CA, September, 1983.

2. Gates, James H., "California's Seismic Design Criteria for Bridges," Journal of the Structural Divisions, ASCE, Vol 102, No. ST12, December, 1976, pp. 2301-2313.

3. "Design Criteria," Adopted by the Sacramento Transit Development Agency, 1982, Sacramento, CA.

4. Greensfedler, R.W., "A map of Maximum Expected Bedrock Acceleration from Earthquakes in California.", California Division of Mines and Geology, Map Sheet 23, Sacramento, CA 1974.

5. "Uniform Building Code," Adopted by the International Conference of Building Officials, 1979.

6. Mayes, Ronald L. and Sharpe, Roland L., "Seismic Design Guidelines for Highway Bridges," Applied Technology Council, Palo Alto, CA and The Federal Highway Administration, Office of Research and Development, Washington, DC, Report No. FHWA/RD-81, October, 1981.

SEISMIC DESIGN OF UNDERGROUND STRUCTURES SOUTHERN CALIFORNIA METRO RAIL PROJECT

J. E. Monsees, *Member ASCE
J. L. Merritt, **Fellow ASCE

This paper dicusses the seismic design of underground structures for the Southern California Rapid Transit District (SCRTD) Metro Rail Project.

I Introduction

The 18.6 miles starter line of the Southern California Rapid Transit District Metro Rail Project is expected to help solve the increasing transportation problems of Los Angeles' high-density urban center -- the Regional Core. With the exception of the yard and shops near Union Station, the entire starter line will be underground in twin tunnels, finished with tunnel liners. This project is unique in the world as it traverses gas-bearing ground, tar sands, and fault zones.

The effects of earthquakes on underground structures may be broadly grouped into two general classes--shaking and faulting. In response to earthquake motion of bedrock (shaking), the soil transmits energy by waves. Seismologists identify various types of earthquake waves, but underground engineers are generally interested in the effects of transverse shear waves, which produce a displacement of the ground transverse to the axis of wave propagation. The orientation of propagation is generally random with respect to any specific structure. Waves propagated parallel to the long axis of a linear structure, such as a subway tunnel, will tend to enforce a corresponding transverse distortion on the structure. Waves traveling at right angles to the structure will tend to move it back and forth longitudinally, and may tend to pull it loose at zones of abrupt transitions in soil conditions, where wave properties may vary. Diagonally impinging waves subject different parts of a linear structure to out-of-phase displacements. This results in a longitudinal compression-rarefaction wave that travels along the structure.

* Principal Technical Associate, Parson Brinckerhoff Quade and Douglas, Inc., and Chief Tunnel Engineer, Metro Rail Transit Consultants, 548 South Spring Street, 7th Floor, Los Angeles, California 90013

** President, Merritt Cases, Inc., P.O. Box 1206, Redlands, California 92373

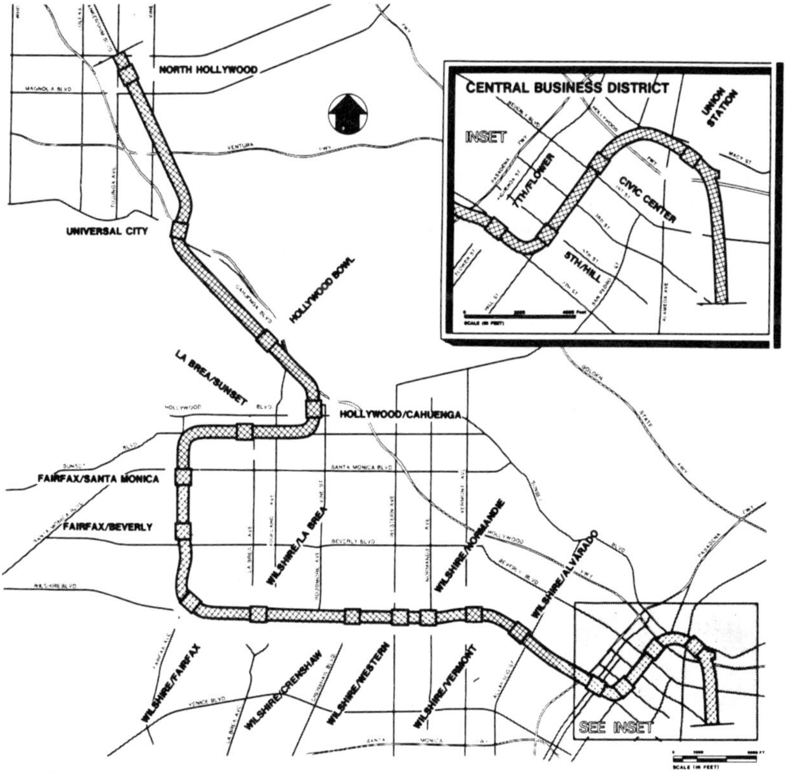

FIGURE 1 ROUTE MAP
SOUTHERN CALIFORNIA METRO RAIL PROJECT

Faulting includes direct primary shearing displacements of bedrock, which may carry through the overburden to the ground surface. Such physical shearing of the rock or soil is generally limited to relatively narrow, seismically active fault zones, which may be identified by geological and seismological surveys.

In general, it is not feasible to design underground structures to restrain major ground displacements. Useful design measures are limited to identifying and avoiding sensitive areas, or if this is not possible, accepting the displacement, localizing and minimizing damage, and providing means to facilitate repairs.

The major contribution to deformations and corresponding stresses in long linear structures such as tunnels is traveling wave effects which can be accounted for by assuming that the tunnel and surrounding soil move together as the wave passes, and that motion from point to point along the route follows the wave pattern and differs from point to point only due to a time lag.

It should also be recognized that although the absolute amplitude of earthquake displacement may be large, this displacement is spread over a long length. The rate of earthquake distortion is generally small, and often within the elastic deformation capacity of the structure. If it can be established that the maximum deformation imposed by the earthquake will not strain the structure beyond the elastic range, no further provisions to resist the deformation are required. If certain parts of the structure are strained into the plastic range, the ductility of such parts must be investigated. If continuity of the structure has been assumed in the design for static loads, the residual effects of plastic distortions may require special consideration.

In the following sections, methods of analyzing the impact of earthquake movements on structural design for underground structures are presented.

II Environments

Design ground motion values for the underground structures are given in the following Table 1 (6).

Where actual surface faulting may be expected to occur, possibly at the Santa Monica and Hollywood Fault, the maximum fault displacement is expected to be 4.7 ft. and 1.6 ft. for the maximum design earthquake (MDE) and operating design earthquake (ODE) respectively (5). The Richter magnitude is approximately 7.0 for MDE and 6.5 for ODE.

TABLE 1
DESIGN EARTHQUAKE PARAMETERS

DESIGN EARTHQUAKE	FOUNDATION CONDITION	DESIGN GROUND MOTION PARAMETERS					
		ACCELERATION (g)		VELOCITY (ft/sec)		DISPLACEMENT (ft)	
		Hor.	Vert.	Hor.	Vert.	Hor.	Vert.
ODE	Soil	0.30	0.20	1.4	1.0	1.6	1.0
	Rock	0.30	0.20	0.8	0.6	0.5	0.3
MDE	Soil	0.60	0.40	3.2	2.1	3.3	2.2
	Rock	0.60	0.40	1.9	1.3	1.0	0.7

The shearing distortion of the ground is given in Figures 2 and 3 which are reproduced here from Converse, (6) with the addition of curves for new alluvium.

Table 2 provides the recommended value in line 3, column 3 for strain to be used in design for earthquake inputs; this table also gives representative values for old alluvium with a shear wave velocity of 1200 fps.

With the postulation that the ground does not lose its integrity during the design earthquake, the basic concept governing the response of underground structures is that the ground is generally stiff compared to the structure and therefore the earthquake deformation of the ground is imposed on the structure, which must conform to this deformation. For very soft soils, interaction between the soil and the structure may be considered, but for any reasonably competent ground this interaction may be neglected, and the structures should be designed to conform to the free soil deformations. Ignoring interaction generally induces larger deformation and strain; thus, it is conservative.

III Stiffness of Underground Structures

Tunnel Lining

The general approach taken in assessing either the static or dynamic behavior of a lined tunnel depends on the relative stiffness of the liner-soil system is conveniently considered as being divided into two separate and distinct types. The first is extensional stiffness, which is a measure of the equal all-around uniform pressure necessary to cause a unit diametral strain of the liner with no change in shape. A measure of the extensional stiffness of the medium relative to that of the liner is given by the compressibility ratio, C, defined by Peck (17):

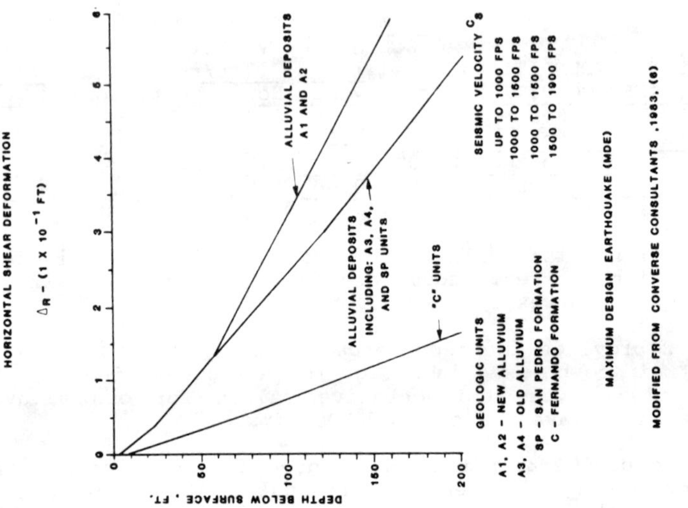

FIGUE 3 HORIZONTAL SHEAR DEFORMATION IN VARIOUS GEOLOGIC UNITS

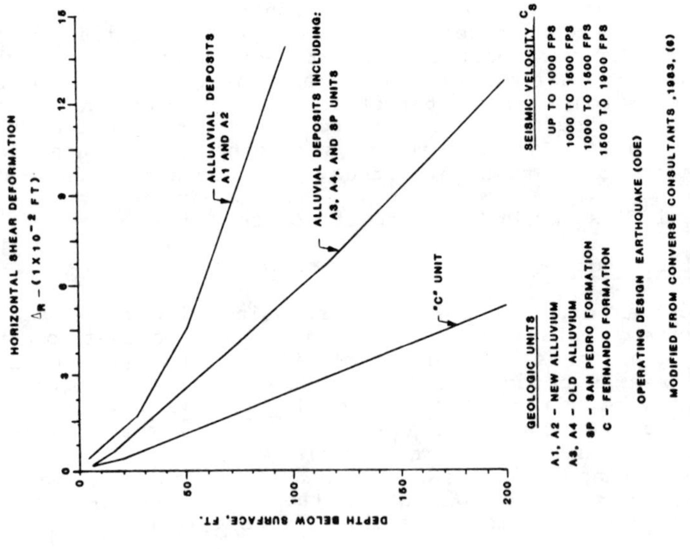

FIGURE 2 HORIZONTAL SHEAR DEFORMATION IN VARIOUS GEOLOGIC UNITS

$$C = \frac{E_m (1 - \nu^2)\ R}{E\ \ t\ (1 + \nu_m)(1 - 2\nu_m)} \qquad 1$$

where E_m = modulus of elasticity of medium
E = modulus of elasticity of liner
R = radius of liner
ν_m = Poisson's ratio of medium
ν = Poisson's ratio of liner
t = average thickness of liner per unit length.

Since the earthquake loadings involve shear waves which primarily change the shape of elements in the groung without significantly changing the average principal stress, the compressibility ratio of the tunnel has very little effect on the behavior of the tunnel.

The flexibility ratio is a measure of the flexural stiffness of the medium relative to that of the lining. The flexural stiffnesses of both the medium and the liner, are measures of the resistance of each to a change in shape (ovaling of the liner) under a state of pure shear. The flexibility ratio, F, is given by the following equation (17):

$$F = \frac{E_m (1 - \nu_1^{\,2})\ R^3}{6E\ I\ \ (1 + \nu_m)} \qquad 2$$

Calculations of liner-medium interaction of various loading conditions (17) show that a liner will behave essentially as a perfectly flexible structure if the flexibility ratio is larger than 20. For the Metro Rail Project, the flexibility ratio obtained using representative properties for the ground and lining is 38 to 50. Thus, the lining is flexible and will conform to the distortions imposed by the medium and the distortions experienced by the liner can be estimated by calculating the distortions for the free field.

Snaking Mode for Tunnel Linings and Stations

Flexibility of the subway structure in plan view, where the structure appears as a long narrow tube or "snake" subjected to the earthquake ground wave motion, has not been defined

Table 2

Selected Principal Effects of Earthquake Generated Waves on Flexible Underground Structures

Wave Type	Propagation Direction Relative to Tunnel Axis (θ)	Effect of Combined Waves Strains Induced in Medium: General Formula[a]	Effect of Combined Waves Strains Induced in Medium: Representative Numerical Value Old Alluvium	Remarks
Dilational (P-) Wave	0°	$\epsilon_p = \pm \frac{v_{max}}{c_{pe}}$	± 0.0017 [b,c] (in/in)	No curvature induced for 0° incidence
Shear (S) Wave	45°	$\epsilon_s = \pm \frac{v_{max}}{2c_{se}} \pm 0.7\ R \frac{a_{max} g}{c_{se}^2}$	± 0.0018 [b]	R = 10' (Assumed) See footnote d.
Recommended Value	Shear Wave at 45°	$\bar{\epsilon}_l = \pm \frac{v_{max}}{2c_{se}} \pm 0.7\ R \frac{a_{max} g}{c_{se}^2}$	± 0.0018 [b]	R = 10' (Assumed) See footnote d.

a Assumes a_{max} and v_{max} are produced only by the wave considered and they occur simultaneously which is physically impossible but an acceptable conservative approximation.

b For old alluvium and MDE; c_s = 1200 fps and c_{se} = 0.8 x 1200 = 960 fps

c $c_p = 2c_s$ for $\nu = 1/3$; this is a better approach for deriving strain in the soil than a measured value in saturated or nearly saturated soil for which c_p approaches that for the water in the interstices. For dry soils $c_p = 2c_s$

d Coefficient includes required trigonometric terms for angle of incidence.

defined mathematically in the literature. However, based on the following considerations it is recommended that the structure be investigated by imposing on the structure, as a design bound, the strains produced in the ground by the earthquake:

a. If the structure exhibits the same flexibility as the ground replaced, assuming the structure strains are equal to the strains in the ground is the exact solution.

b. If the structure is more flexible than the ground, the structure will certainly follow exactly the ground motions so that once again assuming the structure strains equal to the strains in the ground is the exact solution.

c. If the structure is less flexible than the ground, it will attempt to move less than the ground moves in the free field. By moving less than the ground, the structure necessarily is strained less than the ground. Therefore imposing the strains in the ground on the structure is conservative.

Thus, for all cases of structural flexibility the strain in the ground is either equal to or a bound of the strain in the structure, and imposing the strain in the ground on the structure is a conservative assumption.

IV. Earthquake Racking of Stations and Other Shallow Rectangular Framed Structures.

If it can be established that the maximum deformation imposed by the earthquake will not strain the structural frame beyond yield at any point no further provisions to resist the deformation are required. If certain joints are strained into the plastic range, the structure must be checked and redesigned as necessary to assure that no plastic hinge combinations capable of leading to a collapse mechanism can be formed. The general procedure is as follows:

1. Base initial size of members on static design and appropriate strength requirements.

2. Impose earthquake deformation (racking) on the structure using data from Figures 2 and 3 and following the concept shown in Figure 4. These racking deformations induce moments and other internal forces in the structure. These forces, which are treated as the values of Q in Section V, are added to those from the static analysis in accord with Equations 3 and 4. Follow ACI (1, 2) or Uniform Building Code (23) as appropriate

for determining member stiffness. Each racking component (horizontal and vertical) is considered separately, not summed.

3. Evaluate conditions in the structure.

4. Design completed if ultimate conditions in the context of plastic design are not exceeded at any point.

5. Evaluate possible mechanisms, see Figure 5. Conditions such as those in Figure 5-a., are acceptable because a failure (collapse) mechanism has not formed. Conditions such as those in Figure 5-b, are acceptable because collapse is prevented by the surrounding soil. Mechanisms such as 1, 2, 3, 4, or 5 in Figure 5-c, however, would lead to collapse and are, therefore, not acceptable.

6. Check the structure for strain in the longitudinal direction resulting from frictional soil drag.

7. Modify the structure elements as necessary so that an acceptable design results.

V Design of Stations and Other Shallow Rectangular Framed Concrete Structures for dynamic loading conditions.

Dynamic Loading Conditions

Stations will be designed for static loading and then checked for dynamic loading conditions using and the following:

$$U = 1.0\ [1.0\ D + 1.0\ L + 1.0\ H + 1.0\ Q] \quad (3)$$

for the maximum design earthquake,

and

$$U = 0.75\ [1.4D + 1.7L + 1.7H + 1.87\ Q] \quad (4)$$

for the operating design earthquake

where U, D, L and H are as defined by ACI (1, 2)

and

Q are the moments or other internal forces produced by imposing the earthquake distortion on the structure.

Design Details

1. Designer will select the reinforcing steel percentages to avoid brittle behavior.

2. Under static loads, the maximum usable compressive concrete strain ϵ_c for flexure is 0.004 and for axial loading is 0.002 for (8, 9, 10, 12, 21).

3. Redistribution of moments in accordance with ACI 318 -83, Section 8.4 is acceptable (1).

4. Consideration of plastic hinges is acceptable (4, 16, 20).

5. Where the permanent subway structure is designed for "at-rest" earth pressures, no increase in pressures during or subsequent to an earthquake need be considered. Where structures are designed for static active earth pressures, both active and at rest pressures will be used in the above equations for dynamic loading (6, 22).

6. Vertical loads from soil backfill, water in soil, structural components, or other materials, over cut and cover structures will be increased by 20 percent and 40 percent for ODE and MDE respectively.

VI Design of Running Lines

Static Loading Conditions

The current specified concrete segments for running lines have been designed for static loading conditions existing at all soil or soft rock sites* along the alignment. Linings for tunnel sections in the Topanga will be designed by that section's designer.

Dynamic Loading Conditions

The adequacy of these designs to resist the superimposed dynamic loading conditions must be checked following the steps given below. In applying these steps, no load factor is applied to the values of strain or displacement induced by the ODE or MDE. Similarly, no capacity reduction factor or equivalent reduction in strain capability is assigned.

The running line structure is flexible compared to the flexibility of the surrounding medium for distortions in planes perpendicular to the axis of the structure. These distortions are, thus, the same as those of the surrounding medium. Although the running line structure is longitudinally stiffer than the surrounding medium, imposing the motions induced in the medium onto the structure is generally conservative: as discussed in Section III.

---

*Soil or soft rock includes all ground conditions except the Topanga.

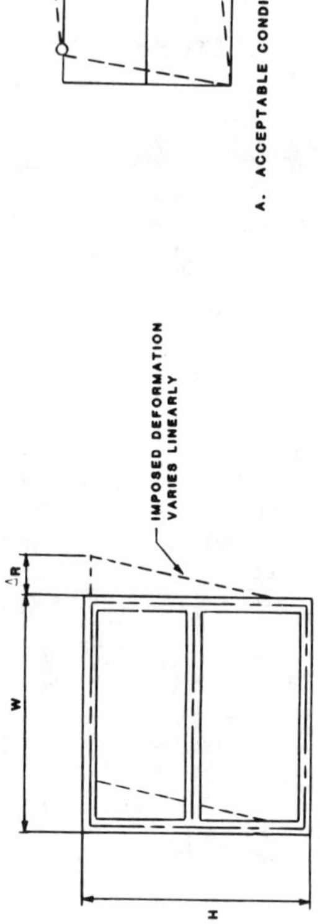

A. IMPOSED HORIZONTAL DEFORMATION

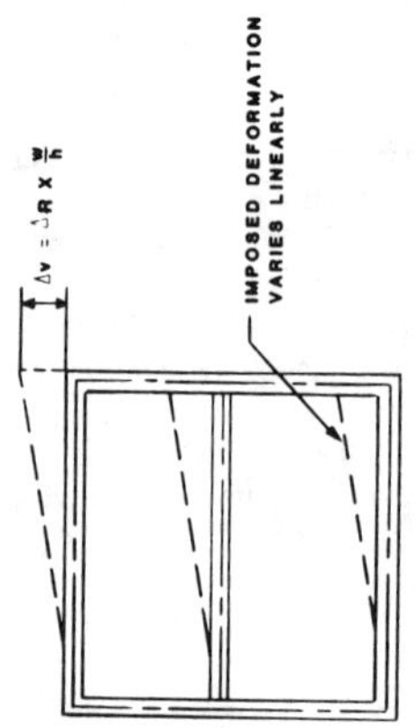

B. IMPOSED VERTICAL DEFORMATION

NOTE: Δ_H (SLOPE OF CURVE TIMES HEIGHT OF STORY OR OF STRUCTURE) FROM FIG. 2 OR FIG. 3

FIGURE 4 APPLICATION OF IMPOSED EARTHQUAKE DEFORMATION

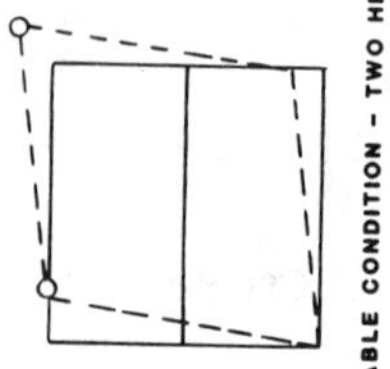

A. ACCEPTABLE CONDITION - TWO HINGES

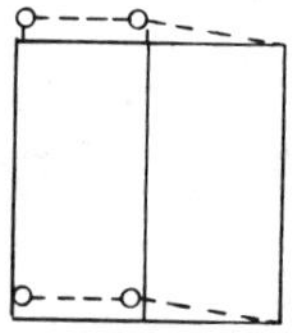

B. ACCEPTABLE CONDITION - FOUR HINGES

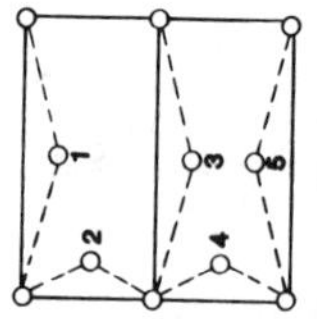

C. UNACCEPTABLE CONDITIONS - THREE HINGES IN ANY MEMBER

FIGURE 5 STRUCTURE MECHANISMS

1. Determine the applicable shear wave velocity c_s for each segment of the running line.

2. Define appropriate value of effective shear wave velocity c_{se}m consistent with the strain level expected. In general unless explicit data are available, c_{se} may be approximately 0.9 c_s for ODE and 0.8 c_s for MDE; in isolated cases where the line is within new alluvium, c_{se} may be approximately 0.75 c_s for ODE and approximately 0.5 c_s for MDE (11).

3. Compute maximum induced longitudinal strains using "recommended value" from Table 2 for the ODE and MDE. The maximum usable compressive strain for this case is 0.002 (8, 9, 10, 16).

4. Assess the longitudinal capabilities of lining for ODE and MDE explicitly considering effects and articulation. Ductile bolts shall be used in all cases.

5. When necessary, modify longitudinal design details to assure no adverse distress for ODE conditions and no collapse for MDE conditions. The minimum reinforcing ratio shall be 0.003.

6. Check strains in plane perpendicular to the axis of the tunnel produced by static design considerations in combination with earthquake racking deformation as discussed below.

Strains produced from static design considerations include uniform compression strains resulting from ground loadings and bending strains produced by horizontal or vertical displacements induced in the lining by construction and by ground loadings (17, 18, 19).

The strain due to racking deformation is that produced by a shear wave with principal distortion in the plane of the tunnel which is perpendicular to the axis of the tunnel. The strains are approximately:

In compression:

$$\epsilon_{rack} = \left(\frac{V_s}{cse}\right) \left[2\left(\frac{t}{R}\right) + \frac{3}{16}\left(\frac{E_m}{E}\right)\left(\frac{t}{R}\right)\right] \qquad 6$$

and in tension:

$$\epsilon_{rack} = 2 \left(\frac{v_s}{c_{se}}\right) \left(\frac{t}{R}\right) \qquad 7$$

where v_s = peak particle velocity produced by earthquake

c_{se} = effective shear wave velocity for the value of v_s (11), see Step 2.

t = thickness of lining

R = radius of lining

E_m = modulus of elasticity of medium

E = modulus of elasticity of lining

7. Compare circumferential strains with allowable values. The maximum usable compresssive strain, ϵ_c, for flexure shall be 0.004 (8, 9, 10, 16).

8. Modify design if required.

VII Closure

The objective of these criteria is to recognize that the effect of the earthquake on underground structures is the imposition of an arbitrary deformation which cannot be changed by strengthening the structure. The structural design goal is therefore to provide sufficient ductility to absorb the imposed deformation without losing the capacity to carry static loads, rather than a criterion of resisting inertial loads at a specified unit stress (3, 7, 13, 14).

VIII Acknowledgements

The authors thank Dr. A. J. Hendron, Jr., University of Illinois, and Mr. T. R. Kuesel, Parsons Brinckerhoff Quade and Douglas for their input to and construction reviews of this work.

REFERENCES

(1) ACI Committee 318 (1983), Building Code Requirements for Reinforced Concrete (ACI 318-83), American Concrete Institute, Detroit.

(2) ACI Committee 318 (1983), Commentary on Building Code Requirements for Reinforced Concrete (ACI 318-83), American Concrete Institute, Detroit.

(3) ATC (1978), "Tentative Provisions for the Development of Seismic Regulations for Buildings. A Cooperative Effort with the Design Professions, Building Code Interests and the Research Community," Applied Technology Council Publication ATC 3-06, NBS Special Publication 510, NSF Publication 78-8.

(4) Blume, J.A. et al. (1961), Design of Multistory Reinforced Concrete Buildings for Earthquake Motions, Portland Cement Association, Chicago.

(5) Bonilla, M.G. (1982), "Evaluation of Potential Surface Faulting and Other Tectonic Deformation," USGS Open - File Report 82-732, Menlo Park, California.

(6) Converse Consultants (1983), "Seismological Investigation & Design Criteria," Southern California Rapid Transit District Metro Rail Project.

(7) Ellingwood, B. et al. (1980), "Development of a Probability Based Load Criterion for American National Standard A58," NBS Special Publication 577, Washington.

(8) Ford, J.S. et al. (1981a), "Behavior of Concrete Columns Under Controlled Lateral Deformation", Journal of the American Concrete Institute, January - February 1981, No. 1, Proceedings V.78, pp. 3-20.

(9) Ford, J.S. et al. (1981b), "Experimental and Analytical Modeling of Unbraced Multipanel Concrete Frames," Journal of the American Concrete Institute, January - February 1981, No. 1, Proceedings V.78, pp. 21-35.

(10) Ford, J.S. et al. (1981c), "Behavior of Unbraced Multipanel Concrete Frames," Journal of the American Concrete Institute, March - April 1981, No. 2, Proceedings V.78, pp. 99-115.

(11) Grant, W.P. and Brown, F.R., Jr. (1981), "Dynamic Behavior of Soil from Field and Laboratory Test, Proceedings, International Conference on Recent Advances in Geotechnical Earthquake Engineering and Soil Dynamics, University of Missouri at Rolla, pp. 591-596.

(12) Hognestad, E. (1951), "A Study of Combined Bending and Axial Load in Reinforced Concrete Members," University of Illinois Bulletin, Volume 49, Number 22, November 1951, Bulletin Series No. 399.

(13) Housner, G.W. and Jennings, P.C. (1982), "Earthquake Design Criteria," Earthquake Engineering Research Institute, Volume Four of a series titled: Engineering Monographs on Earthquake Criteria, Structural Design, and Strong Motion Records.

(14) Newmark, N.M. and Rosenblueth, E. (1971), Fundamentals of Earthquake Engineering, Prentice-Hall, Inc., Englewood Cliffs, N.J.

(15) Nyman, D.J. (1983), Private Communication.

(16) Park, R. and Paulay, T. (1975), Reinforced Concrete Structures, John Wiley & Sons, New York.

(17) Peck, R.B. et al. (1972), "State of the Art of Soft Ground Tunneling," First North American Rapid Excavation and Tunneling Conference, ASCE-AIME, Chicago.

(18) Peck, R.B. 1969, "Deep Excavation and Tunneling in Soft Ground," State of the Art Volume, Seventh International Confernece on Soil Mechanics & Foundations, Mexico City.

(19) Ranken, R.E. et al. (1978), "Analysis of Ground-Liner Interaction for Tunnels," U. S. Department of Transportation Report No. UMTA-IL-06-0043-78-3, Department of Civil Engineering, University of Illinois at Urbana-Champaign Report No. UILU-ENG-78-2021.

(20) Rosenblueth, E. (1980), Design of Earthquake Resis tant Structures, John Wiley & Sons, New York.

(21) Roy, H.E.H. and Sozen, M.A. (1964), "Ductility of Concrete," Flex- ural Mechanics of Reinforced Con crete, Proceedings of the International Symposium, Miami, Florida, 10-12 November 1964, sponsored by ASCE//ACI, pp. 213-235.

(22) Seed, H.B. and Whitman, R.V. (1970), "Design of Earth Retaining Structures for Dynamic Loads," presented at the 1970 Specialty Conference Lateral Stresses in the Ground and Design of Earth-Retaining Struc tures, sponsored by ASCE, 22-24 June 1970 at Cornell University, Ithaca, N.Y., pp. 103-147.

(23) Uniform Building Code (1982), International Conference of Building Officials, Whittier, California.

(24) Wiegel, R.L. et al. (1970), Earthquake Engineering, Prentice Hall, Inc., Englewood Cliffs, N.J.

BASE ISOLATION CONCEPTS FOR SEISMIC BRIDGE RETROFIT

Ronald L Mayes[1], Lindsay R Jones[2], Trevor E Kelly[3], Martin R Button[3], M.ASCE

ABSTRACT

Recent developments in New Zealand on the use of base isolation and energy dissipating mechanisms for the design of new bridges and buildings offer exciting possibilities for the retrofit of existing bridges. The concepts provide retrofit solutions for columns and piers with inadequate strength and for superstructure girders with inadequate support lengths. The superstructure displacements are significantly decreased during seismic excitation and the concepts also provide a force-limiting mechanism for the supporting substructure. The paper discusses the basic concepts and philosophy of base isolation design and presents the application of the concepts on two bridges. The first example is a typical three-span bridge and the impact on base isolation design of different seismic zones is examined. The second example is the CalTrans Sierra Point Overhead seismic retrofit project. This is an 8-span skewed bridge that has recently been retrofitted utilizing base isolation concepts.

INTRODUCTION

The San Fernando earthquake of 1971 demonstrated the potential inadequacy of past design procedures in providing seismically resistant bridges. Since most existing bridges in service in the U.S. today were designed using pre-1971 design criteria, it follows that many of the nation's highway bridges in 37 seismically active states may have insufficient strength to resist seismic loading. Recognizing this problem, CalTrans developed an ongoing program of retrofitting bridges in California [6]. In addition, FHWA funded two major projects that were performed by Applied Technology Council. The first was the development of Seismic Design Guidelines for new bridges [1], and the second was the development of Seismic Retrofit Guidelines for existing bridges [2].

The CalTrans retrofit program primarily involves providing positive connections between superstructure segments and between the superstructure and supporting substructures. This solution aids in the lack of sufficient support lengths for bridge girders. At this time no attempt has been made by CalTrans to strengthen inadequate supporting substructures such as columns and piers primarily because of the expense.

Recent developments in New Zealand on the use of base isolation and energy dissipating mechanisms for the design of new bridges and buildings offer exciting possibilities for the retrofit of existing bridges. The concepts

1 President-Dynamic Isolation Systems, 2855 Telegraph, Berkeley, CA
2 Vice President-Dynamic Isolation Systems, 2855 Telegraph, Berkeley, CA
3 Project Engineer-Dynamic Isolation Systems, 2855 Telegraph, Berkeley, CA

provide retrofit solutions for columns and piers with inadequate strength and for superstructure girders with inadequate support lengths. Superstructure displacements are significantly decreased during seismic excitation and the concepts provide a force-limiting mechanism for the supporting substructure.

The paper presents the application of the devices on two bridges. The first example is a typical three-span bridge and the impact on base isolation design of different seismic zones is examined. The second example is the CalTrans Sierra Point Overhead seismic retrofit project. This is an 8-span skewed bridge that has recently been retrofitted utilizing the base isolation concepts.

SEISMIC RETROFIT CONSIDERATIONS

The Applied Technology Council has recently completed the development of seismic retrofit guidelines for highway bridges [2]. The guidelines are comprehensive and embody several new concepts which are significant departures from existing retrofitting practice. The guidelines include a preliminary screening procedure, methods for evaluating an existing bridge in detail, and potential retrofitting measures for the most common seismic deficiencies. The preliminary screening procedures are used to identify bridges that are the most likely candidates for retrofitting. This procedure rates bridges based on the vulnerability of the structural system, the seismicity of the bridge site, and the importance of the bridge. The methods for evaluating an existing bridge in detail involve, as a first step, the calculation of seismic capacity/demand ratios for each potentially vulnerable bridge component. These capacity/demand ratios are used to assess the consequences of a design earthquake at the bridge site. The results of this assessment can then be used to select the most appropriate retrofitting scheme for the bridge. These retrofitting schemes may be selected from among those presented and discussed in the guidelines. Also included are special design requirements for various retrofitting measures.

The three major problem areas identified in the ATC-6-2 report are:

1. The vulnerability of existing steel rocker and roller type bearings shown in Fig. 1.
2. The inadequate strength and ductility of columns and piers.
3. The inadequate support lengths for girders.

Potential retrofit measures for these problem areas are presented in the ATC-6-2 report. Unfortunately, only a few have been utilized in practice, primarily because of the expense of some of the details that are required to alleviate all three problem areas and especially the inadequate strength and ductility of existing columns. Some of the measures presented in the ATC-6-2 report are summarized as follows:

(a) Vulnerable Bearings: Replacement of bearings should be considered if their failure will result in collapse or loss of function of the superstructure. Types of bearings that have performed poorly in past earthquakes are shown in Fig. 1 and suggested replacements include elastomeric bearings or more sophisticated energy dissipating devices.

(b) Inadequate Strength and Ductility of Columns: Reinforced concrete

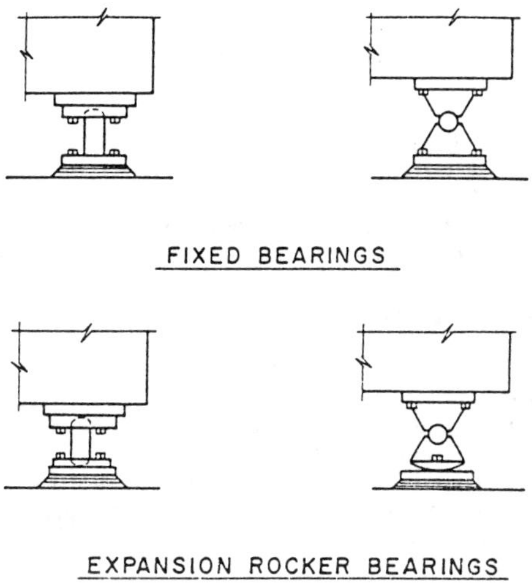

Figure 1 : Seismically Vulnerable Bearings

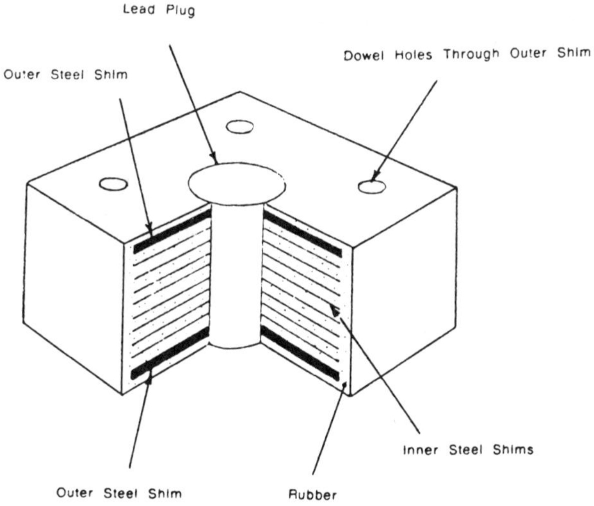

Figure 2 : Typical Lead-Rubber Bearing

columns or piers and the footings to which they are attached form a group of interacting components that are among the most vulnerable to earthquake damage. In quantitatively evaluating the strength of columns and piers, four failure modes should be considered. These are pullout of main reinforcement, splice failures in the main reinforcement, sudden shear failure and loss of flexural capacity due to insufficient confinement. Although some useful research has been performed with respect to bridge column behavior under cyclic loads, the state of the art is such that column evaluation must rely heavily on engineering judgment. Some of the retrofit measures discussed are the use of force-limiting devices, increased transverse reinforcement for confinement of columns, reduced or increased longitudinal reinforcement, infill shear walls between column bents and strengthening of footings.

(c) Inadequate Support Length for Girders: This is the area for which most of the current retrofit measures have been performed in the United States and primarily by CalTrans [6]. These measures have primarily included the provision of positive cable connections between the superstructure and supporting substructure. Measures used by CalTrans include longitudinal joint restrainers, transverse bearing restrainers and vertical motion restrainers. Other concepts presented in the ATC-6-2 report include a bearing seat extension and the use of recently developed bearings and devices which alter the seismic response of the entire structure.

Recent developments in New Zealand on the use of base isolation and energy dissipating mechanisms offer exciting possibilities for the seismic retrofit of bridges. Replacement of vulnerable bearings with force and displacement limiting devices provide an economical solution for each of the three major problem areas discussed above. The concepts and application of these devices is the subject of the remainder of the paper.

BASE ISOLATION CONCEPTS

The isolation of structures from the damaging effects of earthquakes is not a new idea. The first patent for a base isolation scheme was taken out as early as 1909 but until recently very few structures have been built which use these principles. However, new impetus was given to the concept with the successful development of mechanical energy dissipators by the New Zealand Department of Scientific and Industrial Research [3,4]. When used in combination with a flexible isolation device, an energy dissipator can control the response of the structure by limiting displacements and forces and thereby improving seismic performance. As a consequence, several structures have now been designed and built in New Zealand which incorporate base isolation and energy dissipators [5]. Most are bridges, but two buildings and a free-standing chimney have also been protected in this way. Many of the features of base isolation can be used to improve the seismic performance in existing structures. The inclusion of base isolators and energy dissipators when upgrading a structure can significantly enhance its seismic performance.

There are two basic elements in practical base isolation systems as follows:

1. A flexible mounting so that the periods of vibration of the total system are lengthened (shifted) sufficiently to reduce acceleration response, and

2. A damper or energy dissipator so that the relative deflections across the flexible element can be controlled.

An elastomeric bearing with a lead plug inserted into a preformed hole is an efficient means of providing both these elements in a single unit [9]. Bridge structures have for a number of years been supported on elastomeric bearings and thus the concept of the flexible mount is not new. Increasing the thickness of the bearing is a natural step to ensure adequate flexibility and period shift. However, large superstructure displacements will be generated unless steps are taken to control these relative motions. Energy dissipators have been proposed for this purpose [3,4] since they provide both additional elastic stiffness and hysteretic damping.

CHARACTERISTICS OF LEAD-RUBBER BEARINGS

A lead-rubber bearing comprises alternate layers of rubber vulcanized or cemented to thin steel shims, typically with a lead plug placed in the center (Fig. 2). The internal steel shims plus the externally applied vertical load provide confinement for the plug. The rubber in the bearing carries the weight of the structure and the lead plug provides energy dissipation by plastic deformation. Outer steel shims with dowel holes are provided for transfer of lateral force from the structure above to the bearing. Vertical stiffness of the bearing (K_v) and thus the vertical load capacity is inversely proportional to the thickness of individual rubber layers. Hence, bearings have multiple thin rubber layers rather than a single rubber layer. The shear stiffness of the bearing is similarly inversely proportional to the overall thickness of rubber and increases with increased bonded surface area.

The bearing is designed to resist low levels of shear, such as wind and braking loads, elastically with a high initial stiffness (K_u) until a yield level determined by the characteristic strength, Q_d, is reached. The force level, Q_d, is a function of the diameter of the lead plug(s). The post yield stiffness K_d is kept at a minimum to ensure good energy dissipation and low overall structure stiffness during more severe seismic loading. The bi-linear hysteresis curve formed by these two stiffnesses and shown in Fig. 3 has a stable form and large enclosed area, demonstrating the energy absorption properties.

DESIGN PHILOSOPHY

The design philosophy used for incorporating lead-rubber bearings in a structure must consider potential loads imposed on the bearings as follows:

- o Vertical dead and live loads
- o Non-seismic lateral loads arising from temperature changes, creep, wind and other causes
- o Seismic loads

Detailed procedures have been developed [7] to achieve these aims and this section explains the general considerations used to derive the procedures.

Lead-rubber bearings, when installed, form the only load-bearing element supporting the structure above the foundation. Consequently all gravity and lateral loads from the superstructure must be transferred by the bearings. The design

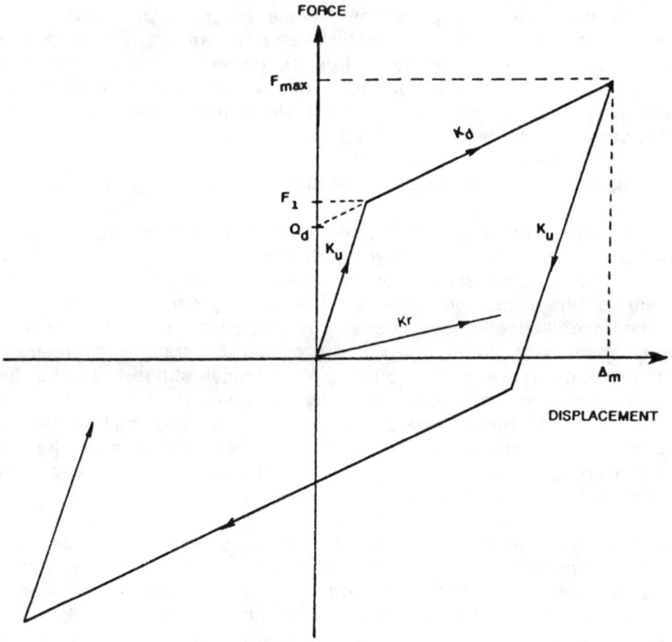

Q_d = Characteristic strength (kips)
F_1 = Yield force (kips)
F_{max} = Maximum force (kips)
K_d = Post-elastic stiffness (kip/inch)
K_u = Elastic (unloading) stiffness (kip/inch) ($=6.5K_d$)
Kr = Elastic stiffness of rubber acting alone (Kip/inch)
Δ_m = Maximum bearing displacment

Figure 3 : Properties of Lead-Rubber Bearing

philosophy for safely transmitting these loads is briefly explained in the following sub-sections.

Vertical Loads: The vertical loads on the bearings arise from the self-weight of the structure and imposed live loads. These loads are normally permanent and to a large degree determinate. The design case for vertical loading is the sum of the dead plus full live loads as for other vertical load carrying elements. Because these loads are considered permanent they form the primary design basis for selection of bearings. Selection of appropriate properties by any design procedure adopted must ensure that the bearings at all times support the maximum vertical load limits with appropriate factors of safety. This is generally achieved by using the rated load approach [10]. Under gravity loads a factor of safety of 3 with respect to the ultimate strain capacity of the rubber is maintained. Vertical load capacity increases with increasing bearing area and decreasing rubber layer thickness. Design charts [7] for square and circular bearings have been developed to quickly determine the appropriate plan dimension.

Non-Seismic Lateral Loads: Lateral loads arising from sources other than earthquakes include those due to wind loads, thermal movements and creep in the structural materials and to vehicle acceleration and braking forces on bridges. The mechanism by which these loads are resisted depends on whether a fast or slow loading rate is involved.

Examples of rapid loading rates would be wind and blast loads on buildings and bridges and braking forces on bridges. These load types have a rate of loading which is fast relative to the period of response of the structure. The design philosophy used for such loads is to set the characteristic yield strength, Q_d, of the bearing at a level such that response will remain elastic. The optimum yield level of the lead-rubber bearings for seismic loads is generally 5% of the structure's weight. If any of these other sources of rapid lateral loading exceed 5% of the structure weight the bearing yield strength should be increased. If these loadings exceed about 10% of the structure weight then the structure is probably not suited for base isolation.

Some sources of loading which are large in magnitude but of short duration and associated with extreme events of small probability can be resisted by permitting the lead to yield as for seismic loads. These could include loads from blasts or impact but would not include severe wind storms.

Typical loadings which are very slow relative to the structural period include thermal loads and material creep. The former may occur over a period of hours, the latter may extend over several years. These loadings are in the form of displacements applied to the bearing and may be of sufficient magnitude to exceed the yield displacement of the lead. The lead will recrystallize and any relaxation in the imposed displacement will cause unloading along the elastic stiffness slope. Any subsequent rapidly applied loads will cause initial elastic response about the new deformed axis. These creep and thermal movements will induce forces in the substructure as they are transmitted through the bearing and may reach the yield level of the plug and slightly above. However, they will not reach the levels of the maximum seismic induced forces and thus will not constitute a governing load case.

If long term creep is expected to be significant, (e.g. in long-span prestressed or post-tensioned buildings or bridges), measures may be incorporated into the design to allow relief of the creep displacements as part of maintainence procedures. This can be done by using slotted holes for the connection of the bearing to the structure and periodically relieving the compression load and returning the bearing to the vertical position. This would be considered only in extreme cases.

Seismic Loads: Installation of lead-rubber bearings beneath a building or bridge structure has two major effects on the bridge response under seismic loads:

1. The effective period at which the structure responds to seismic motions is lengthened considerably compared with the period of the non-isolated structure.
2. The elasto-plastic behavior of the lead plug introduces significant energy dissipation in the form of hysteretic damping.

Detailed design procedures [7] have been developed to quantify these effects and examples of their use follows:

THREE-SPAN BRIDGE EXAMPLE

Fig. 4 shows a three-span steel bridge with 5 main girders. The total number of bearings is 20 with 10 over the piers, each carrying a dead load of 104 kips and a dead plus live load of 124 kips. The abutment bearings carry one-half this load. The detailed steps for designing the lead-rubber bearings for one seismic zone are presented and a comparison of the force and displacement requirements for various seismic zones is then discussed.

1. The maximum dead load plus live load is used with Chart 1-B [7] for circular bearings to obtain the required plan bonded dimension of 13 inches with 3/8 inch internal rubber layers. This bearing has a rated load capacity of 180 kips.
2. The total bonded area of all bearings is then calculated and an average compressive of 588 psi over all bearings is obtained.
3. To obtain the total rubber thickness a design chart for a specific seismic zone and soil type must be used. For this example the CalTrans 0.6g Soil Type S1 design chart is used [7]. It is within this chart that the trade-off in base isolation design between design force versus design displacement can readily be seen. The recommended rubber thickness from the chart varies from 6 to 26 inches. This encompasses a bearing displacement of 6 inches and a base shear force of 0.27W for the 6-inch rubber height where W is the weight of the superstructure. For the 26-inch rubber height, the bearing displacement is 10.4 inches and the base shear force is 0.14W. For design, a rubber height of 8 inches was selected corresponding to an effective period of 1.75 seconds, a force of 0.23W and a maximum displacement of 6.8 inches.
4. Because the abutments carry only one-half the load that the piers carry, it was decided to use lead plugs in only one-half the abutment bearings. To retain symmetry, two plugs are used in each abutment and five in each pier for a total of 14 plugs. To obtain a characteristic strength of 0.05W,

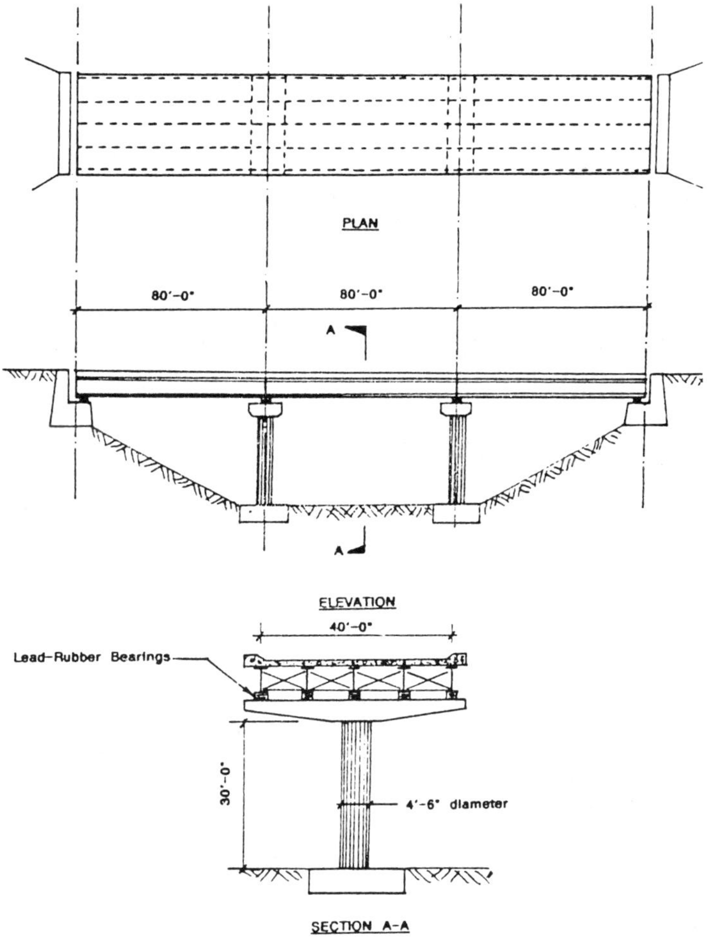

Figure 4 : Bridge Example 1

this required a total strength of (1560 x 0.05) 78 kips, and dividing by 14 provides a required characteristic strength of 5.57 kips per plug. Tabulated properties [7] show that a 2.5" plug provides 5.62 kips, very close to the required value.

If the same bridge were to be located in a moderate seismic zone the same bearing diameter would be used but with a lower rubber thickness. As an example, the ATC-6 seismic design zone [1] with an Aa of 0.30g and soil type S1 Chart 2-B-1 [7] would give an appropriate rubber thickness of 5.00" for the same average compressive stress of 588 psi. This is only 62% the thickness selected for the CalTrans 0.6g design. In this case, the chart provides maximum force levels of 0.14W and corresponding displacements of 2.09". This compares with a force coefficient of 0.23W and 6.84" displacement for the same bridge with 8" rubber thickness and design ground motions corresponding to 0.6g CalTrans.

Table 1 illustrates the variations in force coefficients and displacements as functions of the total rubber thickness and the level of seismicity:

Table 1 : Results for Different Seismic Zones

MAXIMUM GROUND ACCELERATION	T_r = 5"		T_r = 8"		T_r = 11"	
	FORCE COEFF.	MAX. DISP.	FORCE COEFF.	MAX. DISP.	FORCE COEFF.	MAX. DISP.
0.2g-ATC-6	0.10	1.28	0.09	1.60	0.08	1.87
0.3g-ATC-6	0.14	2.09	0.11	2.35	0.10	2.70
0.4g-ATC-6	0.19	2.90	0.14	3.40	0.12	3.71
0.5g-CalTrans	0.26	4.73	0.20	5.20	0.17	5.93
0.6g-CalTrans	0.32	5.73	0.23	6.84	0.21	8.11

The values tabulated above are for the ATC-6 spectral shape for maximum accelerations from 0.2g to 0.4g and for the CalTrans spectra for 0.5g and 0.6g. Therefore, they are not directly comparable for all acceleration levels but do give an indication of the effectiveness of the isolation for varying rubber thicknesses for varying seismic intensity.

It can be seen that the increased rubber thickness is more effective for the higher intensity ground motions. For example, increasing T_r from 5" to 8" reduces forces by only 10% for 0.2g spectra but by 30% when the maximum ground motion is 0.6g.

The table also shows that bearings designed for one seismic intensity are satisfactory for other levels of intensity. For example, the 8" rubber thickness selected for 0.6g ground acceleration also gives good isolation for all intensities from 0.2g to 0.6g. When the 5" rubber thickness selected for the 0.3g region is subjected to higher intensities the force level increases but the base isolation is still effective. In fact, even at 0.6g, double the design earthquake, the rubber strain of (5.73"/5.0") 1.15 is considerably less than the strain levels of 1.50 to which these bearings have been tested.

SIERRA POINT OVERHEAD RETROFIT

The bridge to be retrofitted is a freeway overpass which crosses a secondary road and railroad tracks on Highway 101 just north of the San Francisco Airport. The plan dimensions of the bridge, shown in Fig. 5, are 615 feet long overall and 117 feet wide. The bridge is on a skew of approximately 60 degrees with respect to the highway traffic lanes. The construction is concrete deck slab on a steel superstructure with concrete pier columns 3'-0" diameter and approximately 25 feet high. There are a total of 27 columns, generally arranged in bents of 4 columns, reducing to two or three column bents at the ends of the bridge due to the skew. The foundations are individual pads beneath each column bearing on the underlying sandstone. The steel superstructure consists of a set of longitudinal girders connected to main transverse girders across each column bent. The concrete deck has expansion joints over each pier and at the abutments and at these positions the main girders on one side are free to move longitudinally.

The bridge was selected for retrofit in accordance with the CalTrans program. The substructure, which consisted of cantilever columns, was such that the bridge had a very low seismic capacity, less than one-quarter the site design value of 0.6g maximum ground acceleration. In addition, column detailing was such that no significant ductility could be relied upon. Conventional strengthening measures were considered but proved very expensive and almost impossible to implement due to the presence of railroad tracks adjacent to the columns. An alternative to designing for the maximum forces, that of reducing the forces transmitted into the columns by the maximum earthquake, was then investigated, leading to the decision to assess and eventually adopt the practicality of isolating the superstructure.

Bearing Design Procedures: The form of base isolation was constrained both by physical restrictions imposed by the present bridge bearings and by performance requirements. Bearings were developed to replace existing steel bearings and to meet constraints on height (10" and 11") and plan dimension (less than 19" square). To obtain high vertical load capacity 3/8" rubber layers were adopted. This was a compromise between load capacity and lateral flexibility. Thicker layers have greater lateral load. Thinner layers increase the vertical non-seismic design load but with the penalty of higher lateral stiffness for the same total height, i.e., more of the height is taken up by shims.

These considerations provided two bearing constructions, each 18" square with 1/8" shim plates and 3/8" rubber layers with 3/4" outer shim plates. One bearing was composed of 15 layers for a total height of 8.875" and the other of 17 layers providing 9.875" total height. These bearings had a maximum vertical load capacity of 482 kips, less than the maximum load of 678 kips at some positions. Therefore, 22-inch diameter bearings were also designed, using the same construction as the 18-inch bearings. The circular plan shape allowed bearings of larger size to remain partially within the perimeter of the 19" square existing base plates. A total of 21 bearing locations had total loads (DL + LL) less than 482 kips and therefore were suited for 18" square bearings based on these non-seismic vertical design loads. The remaining 12 bearing locations required 22" circular bearings.

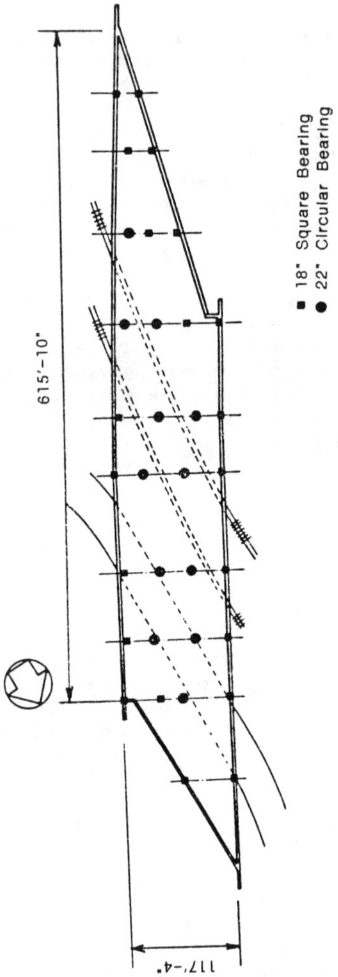

Figure 5 : Plan View of Bridge

Once the basic bearing construction was determined the diameter of the lead plugs and distribution was determined to provide the optimum yield level of 5% of the superstructure weight. The abutment bearings were designed with 6" plugs, the maximum size for their plan dimension. The remainder of the 5% of the weight was achieved by using 3" plugs in all 18" pier bearings and 4" plugs in all 22" bearings. Two abutment bearings had low axial loads, insufficient to adequately confine the lead, and these bearings were specified as elastomeric only.

Prior to a detailed analysis an approximate single-degree-of-freedom estimate of the response quantities was made using the simplified procedures [7]. The effective period was computed to be 1.76 seconds, and for this period the non-linear spectrum provided a maximum acceleration of approximately 0.20g and a maximum displacement of approximately 7.5 inches. These predictions proved accurate as the mean maximum displacement from the 5 time history analyses was 7.0 inches and the mean shear force was 0.20g. Note that this simple check provided only overall response and did not provide detailed checks on the shear force transmitted into each pier column.

Analytical Model: The bridge was analyzed using the ANSR-II computer program [8] for the three-dimensional analysis of non-linear structures. This program permits the bilinear hysteresis curve of the lead-rubber bearings to be included. The model provided the response of the bridge to both longitudinal and transverse earthquake motions. Each column and each bearing was modelled as a three-dimensional beam element, with the columns remaining elastic and the bearing elements having bending properties and yield moments equivalent to the shear stiffness and characteristic strength of the lead-rubber bearings. The earthquake input appropriate for the site was the CalTrans 0.6g maximum acceleration on a rock site. Time histories were obtained for the analysis by frequency scaling a suite of recorded and artificial time histories to closely match the "target" spectrum. The starting time histories were selected as El Centro, 1940 N-S, Taft, 1952 S69E, Olympia 1949 NO4W and Caltech artificial records A1 and A2.

Results of Analysis: The results from the series of 5 time history analyses are presented in Table 2. Tabulated values include strains, forces and force ratios for all 5 records plus the mean value of each quantity.

Table 2 : Analysis Results for Sierra Point

RESPONSE QUANTITY	EARTHQUAKE (Scaled to 0.6g CalTrans)					MEAN
	El Centro	Taft	Olympia	Caltech A1	Caltech A2	VALUE
Max. Displ. (ins.)	6.44	9.73	7.20	3.71	7.77	6.97
Max. Bearing Strain	1.14	1.73	1.28	0.66	1.38	1.24
Total Shear Force (kips)	1728	2388	1763	1116	1870	
V/W	0.20g	0.27g	0.20g	0.13g	0.21g	0.20g
Max. Col. Shear (kips)	53.7	74.2	56.0	34.13	58.9	55.4
V_{max}/V_n	1.06	1.49	1.10	0.69	1.15	1.10
V_{max}/V_o	0.65	0.92	0.68	0.42	0.71	0.68

The maximum shear forces in each column were divided by the nominal capacity, V_n, and the overstrength capacity, V_o. The latter quantity is based on the ATC-6-2 retrofit studies [2] which conclude that columns will perform adequately provided the elastic moment to which they are subjected does not exceed 1.625 times the nominal moment capacity. The mean nominal force ratio was 1.10 and the mean overstrength ratio was 0.68. All records produced overstrength ratios less than unity. Maximum bearing strains were 114%, 128% and 138% for the El Centro, Olympia and Caltech A2 based time histories, respectively. The upper limit used for new design is generally accepted to be 100% but values up to 150% are accepted for retrofit for a limited number of cycles. In this application a maximum of 4 cycles exceeded 100% strain.

The results of the analyses demonstrated that the bearing layout and construction selected provided satisfactory performance for a retrofit situation. Detailing for the installation of the bearings into the structure was then performed.

CONCLUSIONS

Clearly, the seismic safety of existing bridges is of major concern and recent efforts have been directed towards providing guidelines to solving the problem. Failures in past earthquakes have indicated three major problem areas with existing bridges. The first is the vulnerability of existing rocker and roller type bearings and this is coupled with inadequate support lengths for superstructure girders. The third problem area is inadequate strength and ductility capacity of supporting substructures.

The paper has presented the application of a recently developed lead-rubber bearing for the retrofit evaluation of two bridges. The bearings provide both base isolation and energy dissipation characteristics that decrease superstructure displacements and provide a force-limiting mechanism for the supporting substructure. In the three-span bridge example the impact of different seismic zones was assessed and it was shown that bearings designed for one seismic intensity are satisfactory for other levels of intensity. The Sierra Point Overhead bridge in its present as-built condition is capable of resisting less than 25% of the CalTrans design ground motion for the site which is close to a major fault. Incorporation of the lead-rubber bearings between the superstructure and the columns and abutments reduces the forces transmitted into the weak columns by a factor of 5, thus enabling the bridge columns to respond elastically to the design ground motion. These examples illustrate that the recently developed base isolation and energy dissipating concepts embodied in the lead-rubber bearing provide an effective and economical solution for the three major problems of seismic bridge retrofit.

REFERENCES

1. Applied Technology Council, "Seismic Design Guidelines for Highway Bridges," ATC-6 Report, Palo Alto, CA., 1981.

2. Applied Technology Council, "Seismic Retrofitting Guidelines for Highway Bridges," ATC-6-2 Report, Palo Alto, CA., 1984.

3. Blakeley, R.W.G., "Analysis and Design of Bridges Incorporating Mechanical Energy Dissipating Devices for Earthquake Resistance," Proceedings of a Workshop on Earthquake Resistance of Highway Bridges, Applied Technology Council, Report No. ATC-6-1, Nov. 1979.

4. Blakeley, R.W.G., et al., "Recommendations for the Design and Construction of Base Isolated Structures," Bulletin New Zealand National Society for Earthquake Engineering, Vol. 12, No. 2, 1979.

5. Buckle, I.G., "The Use of Energy Dissipators and Base Isolation in the Design and Retrofit of Bridges and Buildings," Proceedings of 51st Structural Engineers Association of California Convention, October, 1982.

6. Degenkolb, O.H., "Retrofitting of Existing Highway Bridges Subjected to Seismic Loading - Practical Considerations," Proceedings of Workshop on Earthquake Resistance of Highway Bridges, ATC-6-1 Report, Palo Alto, CA., 1979.

7. Dynamic Isolation Systems, Inc., "Seismic Base Isolation Using Lead-Rubber Bearings - Design Procedures for Bridges," Berkeley, CA., 1983.

8. Mondkar, D.P. and Powell, G.H., "ANSR-II Analysis of Nonlinear Structural Response User's Manual," University of California, CA., Report No. UCB/EERC 79/17, July 1979.

9. Robinson, W.H., "Lead-Rubber Hysteretic Bearings Suitable for Protecting Structures During Earthquakes," Journal of Earthquake Engineering and Structural Dynamics, Vol. 10, 1982.

10. Stanton, J.F., and Roeder, C.W., "Elastomeric Bearing Design Construction and Materials," NCHRP Report 248, Transportation Research Board, 1982.

SEISMIC RISK AND DECISION ANALYSIS OF LIFELINE SYSTEMS

by Ronald T. Eguchi,* A.M. ASCE

ABSTRACT

This paper provides general guidelines for performing lifeline seismic risk and decision analyses. Relevant input models are introduced as they support the performance of component and systems analyses. To illustrate the types and ranges of output that can be obtained from these analyses, a number of examples are provided. The present paper is based on an earlier publication prepared by the Seismic Risk Committee of TCLEE [6].

INTRODUCTION

The design and analysis of lifeline systems to earthquake hazards have received considerable attention since the 1971 San Fernando earthquake. In that earthquake, systems (both with and without aseismic design features) were subjected to strong ground shaking and differential earth movements. The behavior of lifelines in that earthquake showed up some glaring hazards [20, 42] and demonstrated the need for new design approaches and requirements for lifeline systems.

Since that earthquake, a number of studies in the area of lifeline risk and decision analysis have been undertaken. A large portion of these studies have emphasized the importance of system analysis techniques. Efficient procedures are presented to enable potential users to identify vulnerable components in a system, as well as to estimate expected system performance. The remaining studies have focused on assessing the earthquake performance of discrete components. In particular, detailed analytical and empirical performance models have been developed for such components as underground pipes and liquid storage tanks. All of the above studies have contributed to the state of the practice as well as state of the art of lifeline earthquake engineering. A number of conference publications contain summaries of many of the above studies [3, 4, 5, 6, 7, 8, 9].

The purpose of this paper is to provide general guidelines for performing lifeline seismic risk and decision analyses. In the introductory part of this paper, a general overview of procedures is presented, with preliminary discussions of quantitative measures of network performance, relevant models needed, the format of analyses, and possible applications. In the following section on component vulnerability, a discussion of procedures for analyzing the seismic vulnerability of site-specific components is given. In the third main section

*Agbabian Associates, 250 N. Nash Street, El Segundo, California 90245

on performance criteria, alternative means for evaluating networks or components are outlined. In the fourth main section on system vulnerability, methods for analyzing networks are indicated. In the final major section, steps in decision making are indicated.

Quantitative Measures of Network Performance

Network theory permits us to identify the following quantitative measures of network performance. They are presented in increasing order of complexity, and their advantages or shortcomings are briefly described. Representative quantitative measures of network performance are connectivity, maximum flow, serviceability, and economic loss.

In terms of network mathematics, which is in general, combinatorial algebra [27], the connectivity of a network is readily calculated as a binary variable. However, it is usually difficult to identify a meaningful connectivity test. In a typical example, one can ask whether the water supply remains connected to any particular node, or to some selected set of nodes. However, there is no further quantification of the effect.

In a capacitated network, the maximum flow from any source to any sink can be readily calculated. In a water system example, an artificial "super-source" would feed links which are then real sources in the system. Similarly, an artificial "super-sink" is fed by numerous links which are the real nodal consumption constraints. This calculation gives a better measure of system performance, and can be roughly interpreted as bounds on the economic loss [14, 23, 31, 39].

Network serviceability may be measured by "percentage of population serviced." It is a multidimensional measure of connectivity, in one sense, or it can deal directly with the amounts available at each population node. Since the analyst defines this measure, it can usually be made tractable analytically. Because of its extra "dimensions," it is a good relative measure of overall lifeline performance [15, 22, 33, 44, 55].

The most ambitious measure of network performance is one which measures the economic loss directly from the shortfall of supplies to users. This requires some explicit modeling of user preference, such as embodied in a demand function. It has been attempted for water [14, 31] and transportation [32] systems, but it may be too involved for most applications.

Relevant Models

The relevant models for a lifeline seismic risk analysis are shown in Figure 1. They include seismic hazard, component vulnerability, network, performance criteria, and final system risk models.

The seismic hazard model expresses the likelihood of exceeding various levels of earthquake intensity, within some time period. Intensity measures should characterize the effects from seismic shaking,

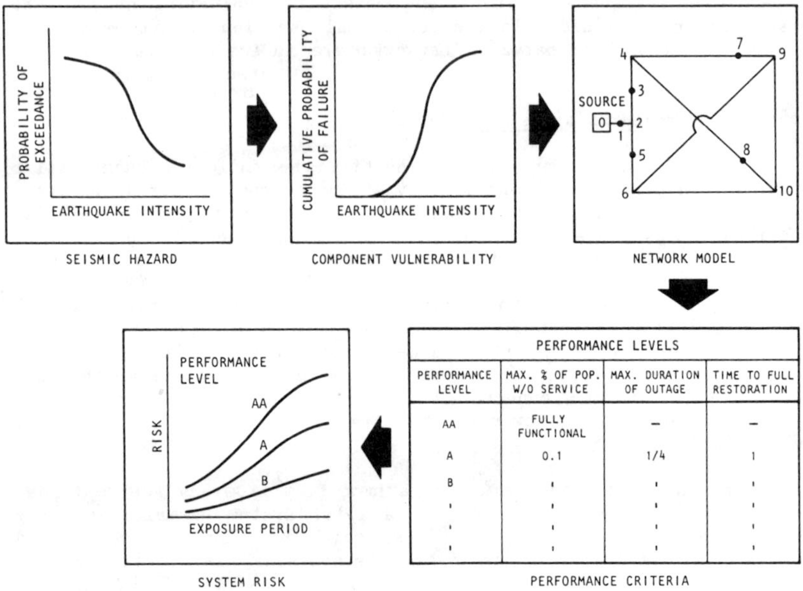

PERFORMANCE LEVELS			
PERFORMANCE LEVEL	MAX. % OF POP. W/O SERVICE	MAX. DURATION OF OUTAGE	TIME TO FULL RESTORATION
AA	FULLY FUNCTIONAL	–	–
A	0.1	1/4	1
B	'	'	'
'	'	'	'
'	'	'	'
'	'	'	'

FIGURE 1. RELEVANT MODELS FOR A LIFELINE RISK ANALYSIS

fault rupture, liquefaction, and slope movement. Assessing the likelihood of these hazards, excluding fault rupture, requires an estimate of the level and duration of ground shaking at the site of interest. Fault rupture depends more on the size and depth of the earthquake.

Published seismic hazard maps, e.g., Algermissen and Perkins [2], are not completely applicable to lifeline risk analyses. Those maps are designed to estimate average effects at a site over many potential events rather than the effects at many sites from one event. Therefore, what is required is a map of the relevant seismic parameters for every earthquake that can be expected to damage the system, together with an assessment of their probabilities of occurrence. Guidelines for estimating the seismic hazard of an area spanned by a lifeline system are provided by ASCE [7].

Once the earthquake hazard is specified, the seismic performance of discrete components must be assessed. Required in these assessments are seismic vulnerability models that relate cumulative probability of failure to some measure of earthquake intensity. Seismic vulnerability models should reflect those earthquake hazard effects that contribute most to the damage of a component. For instance, recent studies have generally shown that the performance of an underground pipe during an earthquake is strongly correlated to whether the pipe is subjected to differential ground movements or not [23, 41, 46]. Vibratory effects appear especially significant only when pipes are in already degraded states, e.g., corroded. Similar assessments of failure should also be made for other components.

The network model, represented by a mathematical expression, relates component performance to system performance. In general, this model is made up of elements in series, in parallel, or in a series/parallel combination. Elements in series, for example, represent a part of the system where the proper operation of all elements is required. Elements in parallel, on the other hand, often make up a part of the system that exhibits redundant properties; proper operation may depend on only part of the system functioning. The expressions for parallel and series elements follow directly from standard Boolean techniques and are essential tools in assessing the overall performance of a system.

Depending upon the objectives of the analysis, different sets of elements may be prescribed in the network model. For instance, an emphasis on operational performance may lead to a different set of system components than that prescribed by vulnerability analyses. Regulator valves, for example, are essential elements in water hydraulic analyses. (The importance of operational analyses in water system vulnerability studies has been discussed by several investigators [24, 56].) However, with regard to earthquake vulnerability studies, these components are sometimes omitted because of their lower relative level of seismic vulnerability.

In order to assess the performance of a lifeline system during an earthquake, it is necessary that performance be measured against nominal standards. As will be discussed in a later section, performance criteria for lifelines may be defined in terms of minimum service requirements for different postearthquake periods. Performance standards will generally consider such factors as the maximum percent of population losing service, maximum duration of outage, and time to full restoration, in addition to any safety-related factors. Specific examples are considered later in the paper.

The final system risk model expresses the likelihood of not meeting various levels of performance, within some time period. Normally, this risk model is presented in the form of a risk curve that increases with exposure period and higher performance standards or requirements (see Fig. 1). Risk, in this case, may be defined by the probability of not being able to serve at least 10% of the serviced population, or the probability that the system duration will exceed 14 days, or the probability that the system will not be restored to full operation within 90 days. Depending upon the criticality of the system, these standards could vary considerably. Note also that a reliability estimate of the system can be made by subtracting the risk value from 1.

Figure 2 illustrates some of the considerations in a lifeline seismic decision analysis as applied to design. The basic method considers balancing the costs of providing increased seismic resistance with costs associated with expected damage to the system. Normally, optimality criteria are established (which may consider safety and social consequences), and a corresponding optimum strategy computed

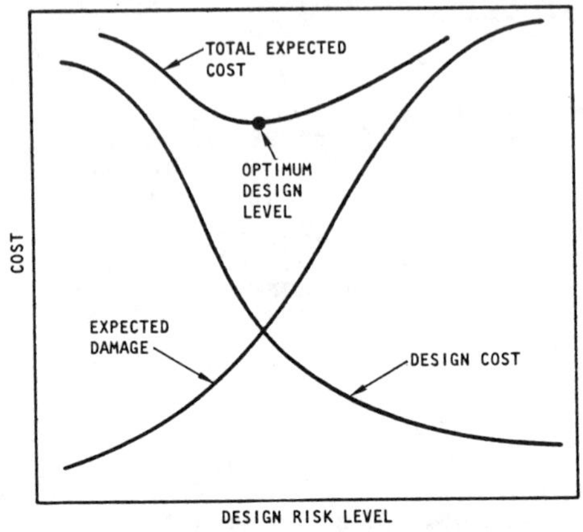

FIGURE 2. LIFELINE DECISION ANALYSIS

[34]. Figure 2 shows the optimum design risk level as the lowest point on the total expected cost curve. That is, further increase in the design risk level to account for added seismic resistance is not warranted from an economic standpoint since the expected damage computed will be less than the added design costs. Other factors related to safety, however, can supersede the results of a pure economic analysis. The final section of this paper discusses some methods related to optimum design strategies for lifeline systems.

Analysis Format

There are typically several ways in which to evaluate the performance of a lifeline system during an earthquake. One way is to assess the performance of the system in one or several earthquakes, i.e., a multiscenario approach. Another way is to assess the performance of the system in all possible earthquakes expected to occur within some period of time, i.e., a total risk approach. Each approach is directed to meeting different sets of objectives.

The multiscenario approach has been used to test the performance of a lifeline system in catastrophic events [24]. Results from these analyses tend to be used to establish upper bound performance requirements on the system. For this reason, they are useful for emergency and recovery planning purposes.

The total risk approach has been used to establish the average performance of a system from a set of earthquakes expected to occur within some period of time. Results from this type of analysis have been used to test the efficacy of various design and retrofit strategies [39].

Applications

Lifeline risk and decision analysis methods have numerous applications in the design and analysis of utility and transportation systems. In general, these methods have not been widely applied primarily because they lacked some of the models described previously. However, recent studies have begun to focus on these models, and their development will facilitate the usage of these methods. Some of the potential applications of these methods are described below.

Lifeline risk analysis methods have been used to assess vulnerable components in a system. This information, in turn, has been used to assess system vulnerability. The results from this analysis can be used in vulnerability assessments, in the retrofit of existing facilities, in the assessment of insurance requirements, and in emergency response planning.

Decision analysis methods, on the other hand, have their greatest potential application in the design of new facilities and/or the retrofit of existing ones. Reliability or service requirements can be assessed for critical components and subsystems. Attainment of these requirements can be derived systematically and optimally through use of these methods.

COMPONENT VULNERABILITY

Seismic vulnerability models relate cumulative probability of failure to earthquake intensity. As indicated earlier, these models should reflect those earthquake hazard effects that contribute most to the damage of a component. In general, aboveground components tend to be vulnerable to vibratory effects caused by strong ground shaking, while underground components tend to be vulnerable to permanent ground displacements caused by tectonic and nontectonic ground movements.

Seismic vulnerability models are typically qualitative, empirical, or analytical. Qualitative models generally indicate whether a component is susceptible to failure or damage from a particular hazard effect. Degree of vulnerability or hazard level is normally not addressed. Empirical models attempt to quantify degree of vulnerability by fitting statistical distributions about limited data bases. In practice, these types of models are the most common. Analytical models provide the most detailed level of representing seismic vulnerability. These models relate component performance to those factors which directly influence the response of a component. The application of these models, however, has been limited primarily because of large information requirements and because they are difficult to apply.

Component Types

Those components of special concern are those which have operational importance or whose damage may create direct hazards to life, safety, and health, and also those that are vulnerable to failure from earthquake shaking or related hazards. For various lifeline networks, Table 1 lists components that have been given special attention. Other

TABLE 1. A SELECTED LIST OF LIFELINE NETWORK COMPONENTS OF SPECIAL CONCERN IN EARTHQUAKE HAZARD REDUCTION STUDIES [62]

Electric Power	Natural Gas	Petroleum	Communications	Transportation	Culinary Water	Sewage Disposal
Nuclear power plants	Pipelines	Refineries	Telephone control stations	Bridge structures	Dams	Pumping plants
Hydroelectric generating facilities	Service connections	Tank farms	Telephone equipment buildings	Tunnels	Treatment plants	Pipelines
Oil or gas storage tanks at steam generating facilities	Gas storage facilities, including critical liquefied natural gas terminals	Oil field facilities	Microwave equipment buildings	Railway tracks and roadbeds	Pumping plants	Treatment plants
Cooling towers	Compressor stations	Pumping stations	Emergency broadcast facilities	Runways	Pipelines	Offshore facilities
Control rooms	Control rooms	Valve locations	Emergency communications centers and equipment	Air traffic control towers	Wells	
AC-DC converter stations		Pipelines	Radio towers, microwave antennas, and repeater installations	Port facilities	Storage facilities	
Bulk substations or switchyards		Monitoring systems		Railway communications facilities	Tunnels	
Backup generators (for several lifelines)				Subways	Canals	
					Aqueducts	

TABLE 2. RELATIVE VULNERABILITIES FOR UNDERGROUND PIPE [23]

Pipe Type	Ground Shaking, MMI		Liquefaction, Lurching	Landsliding	Fault Rupture, displacement, in.	
	8	10			10	100
WSAWJ (modern)	L	L	L	L	L	L
WSAWJ (premodern)	M	M	M	M	M	M
WSGWJ	M	H	H	H	H	H
WSCJ	M	M	H	H	H	H
CI	M	M	H	H	H	H
AC	M	M	H	H	H	H

Pipe Types:

- WSAWJ Welded steel with arc-welded joints
- WSGWJ Welded steel with gas-welded joints
- WSCJ Welded steel with caulked joints
- CI Cast iron
- AC Asbestos cement

Vulnerability Levels:

- Low < 0.03 breaks per 1000 ft
- Moderate 0.03 to 0.3 breaks per 1000 ft
- High > 0.3 breaks per 1000 ft

significant components could also be mentioned, some of which, like electric power transmission towers, can fail, although their design often incorporates earthquake resistance and their siting often includes concern for ground failure, rock slides, and the like. Some facilities in Table 1, such as refineries or hydroelectric generating facilities, could be given much more detailed attention. Other facilities, such as oil pipelines, nuclear power plants, and the recent liquefied natural gas (LNG) receiving terminal off the California coast, have received much attention [10,28]

Other distinctions that are made between components include whether the component is located above or belowground, and whether it has been seismically designed or not.

Performance States

Component performance states can be modeled as binary, multistate, or continuous variables. Binary states often define dichotomous conditions such as damaged vs. undamaged, on vs. off, functional vs. nonfunctional, and open vs. closed. Multistate definitions expand upon the previous by including various levels of performance or damage, e.g., no damage, light damage, moderate damage, heavy damage, total destruction/ inoperative. Continuous vulnerability models represent a continuum of possible states for various levels of earthquake intensity.

While use of continuous performance states is desirable for most lifeline components, in most cases it is not practical. For most components, dichotomous modeling may be adequate. For example, in most reliability studies, all that is required with regard to component modeling is an assessment of functionality, i.e., is the component able to function or not. For more complicated nodes and/or links (e.g., power generating facilities, purification facilities, electric power substations), multistate modeling offers a reasonable approximation to continuous performance state modeling.

The definition of a performance state will depend upon the objective of the analysis being performed. For example, in an economic analysis of expected damage, performance states will be related to ranges of dollar loss [14, 31, 32]. For serviceability studies, performance states will be defined by whether components are functional or not and can provide service [24, 39]. Thus, depending upon the analysis, performance states will differ.

Vulnerability Model Types

Several types of vulnerability models exist, and generally they fall into one of three classes: qualitative, empirical, and analytical.

The simplest vulnerability model consists of qualitative statements of vulnerability [35, 37, 65]. For each component, earthquake intensity characteristics are broken down into various hazard groups (e.g., surface fault movement, soil failure, landsliding, tsunami) and vulnerability levels are assigned. Comparative assignments of component vulnerability are thereby made. At this level of analysis, though, "success" and "failure" of components need to be carefully defined, and may be alternatively defined for various purposes. Damage to an aqueduct or to a chemical building may not impede operations, and temporary power loss at a pumping plant may be comparatively insignificant if the backup generator merely needs to be turned on. An example of a qualitative vulnerability model for underground pipeline components is presented in Table 2.

Empirical vulnerability models, based primarily on statistical observations from past earthquakes, have been used extensively in lifeline risk analysis. All such modeling depends on the correlation of percent damage or failure rate to some measure of earthquake intensity, [23, 26, 36]. This type of modeling is usually generic in the sense that facilities are classified by type. It is possible, for instance, that specific investigations of a given component indicate that it is either more vulnerable or less vulnerable than indicated by the generic curve. However, for most systems analysis purposes, these models are adequate, at least to indicate where more refined analyses are desirable. Examples of empirical vulnerability models for underground pipes are presented in Figure 3.

Analytical models provide the most detailed level of representing the earthquake vulnerability of lifeline components. Tested and further developed through empirical and experimental data, these models allow one to relate damage to factors that directly influence the response of the components, such as rigidity or expected energy dissipation. Typical pipeline response models use such parameters as pipe/fault line orientations, effective pipe free lengths, joint types, and surrounding soil parameters [43, 45, 46, 49, 54, 62]. Much work has been done to develop analytic vulnerability models for buildings [13, 16, 17, 29]. Such analytic models have been based on detailed structural models, i.e., beam and column elements, diaphragms, shear walls, etc. The application of these models are often limited to only the most critical components, because of their larger information and analysis requirements.

SYSTEM AND COMPONENT PERFORMANCE CRITERIA

Since the early work of Duke and Matthiesen [19] and Duke and Moran [21] very little has been done in the area of system and component performance criteria development. Perhaps the biggest reason for this slow development has been the fact that performance requirements will tend to differ from one system to another. Generally, a number of factors influence the level of performance required: operator's requirements, budget constraints, uniqueness of particular facilities, safety importance, public or service requirements, etc. As a result, perhaps the most important contribution to the development of this area would be the definition of useful performance measures. The present subsection attempts to summarize the limited work published in this area. Early guidelines, as presented by the above investigations, are given as examples.

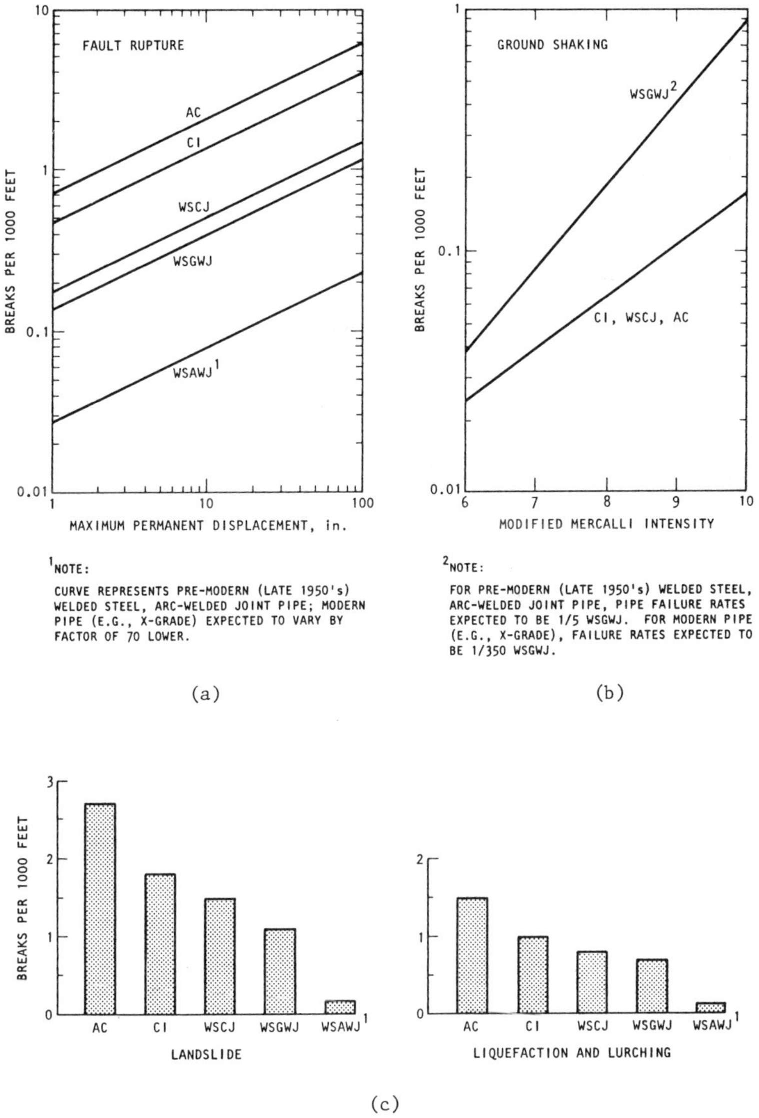

FIGURE 3. EARTHQUAKE VULNERABILITY RELATIONSHIPS FOR UNDERGROUND PIPE (See pipe code in Table 2 for pipe definitions) [23]

Component Performance Criteria

Performance standards for lifeline components have by and large been absent in the literature. To date, only three major reports regarding component performance standards have been published.

First, the Agbabian Associates 1973 report, "Procedures and Criteria for Increasing the Earthquake Resistance Level of Electrical Substations and Special Installations" contains a comprehensive seismic analysis of existing substations, one AC-DC converter station, and one control center in the Bonneville Power Administration transmission network [1]. In their design modification recommendation, the authors produce general operational priorities for substation components. Sometimes systems analyses may not be needed for some components, e.g., circuit breakers are of such operational importance that extensive seismic research has been undertaken and its findings have been implemented to safeguard circuit breakers against seismic failure [61].

Second, in lieu of roadway analyses, criteria have been developed for estimating criticality of highway bridge structures in reports first by Longinow et al. [38] and later by Robinson et al. [48]. The most critical bridge structures are those that are survival related, or critical and nonbypassable, or those carrying over 80,000 vehicles on the average per day. Criticality ratings (ranging from 0 to 3) are contrasted to structural ratings (ranging from 0 to 3) to yield a decision tree analysis procedure for retrofitting highway bridge structures.

Third, Environmental Quality Systems drafted "Earthquake Design Criteria for Water Supply and Wastewater Systems" in 1980, a thorough document that contains recommended design considerations and also criteria for water and sewage facilities, especially treatment plants [25]. In the development of such criteria, seismic risk analyses are mentioned only in passing. In place of such analysis, a functional analysis of treatment plant components is used. On a scale from 1 to 7, with 1 being most critically needed, the following required availability ratings are given to selected water treatment plant components:

Filtration tank	1 or 2
Influence pipes, valves	1 or 2
Air compressors	7
Backwash water trough	5 or 6

Such ratings incorporate emergency response needs, including prioritization of components based on how quickly they need to be restored after a disaster.

System Performance Criteria

With regard to system performance, one criterion was suggested by Campbell et al. [15]: the availability to the public of the use of a functioning lifeline. Another criterion would be the need for a safe, functioning lifeline. These two criteria are overlapping and, of course, are qualitative.

The level of performance has to be expressed quantitatively. One means of judging performance could be the number of people suffering from an outage, the extent of the outage, and the time required to restore the lifeline fully. This concept of performance is given in Table 3. An alternate set of performance levels based on areal distribution of outages is presented at the bottom of Table 4.

Table 4 also provides an example of possible acceptable performance criteria for various lifeline systems under seismic loading. This is illustrated for two levels of intensity of vibratory motion and surface faulting, using the Modified Mercalli Intensity scale, and applies to both new and upgraded facilities.

Table 4 is an aid to quantifying full and partial outages. Using sewage collection facilities as an example, one can see how system performance might be developed:

1. In a major earthquake the sewage collection facilities should be able to perform to at least a C level.

2. Level C calls for 50% or less of the high intensity area to be without service for one week, for 20% or less to be without service for 1 month, and full restoration of service within 3 months.

It is significant that Table 3 does not incorporate the notion of acceptable performance whereas Table 4 does incorporate this aspect. The effect on people in Table 3 could be added (approximately) to the considerations of Table 4.

In addition to the above performance standards, recent studies on water system performance have indicated that more quantitative measures, such as minimum flow rates and pressures, may be used to set standards [22, 23, 55]. These measures require that operational analyses, such as hydraulic analyses, be performed in addition to vulnerability studies.

Reliability Criteria

Reliability is typically defined as the probability that a lifeline system will meet a set of minimum performance standards. This reliability criterion has three parts. One part is the acceptable reliability, essentially a standard quantity that sets a lower limit of reliability. A second part is the exposure period - the length of time over which the system is subjected to the threat of an earthquake. The exposure period may represent the economic life of the system, or the time between major upgradings of system components. The third element is the acceptable performance level, which sets a maximum limit for the magnitude and duration of an outage.

Reliability criteria used in analysis and design permit the engineer to compare the cost effectiveness of different mitigation and design alternatives.

TABLE 3. SUGGESTED PERFORMANCE LEVELS FOR LIFELINE SYSTEMS

Performance Level	Maximum Percent of Serviced Population Without Service	Maximum Duration of Outage, in Days	Time to Full Restoration, in Days
AA	Fully functional		
A	0.1	1/4	1
B	0.5	1	7
C	1	3	14
D	2	7	30
E	10	14	90
F	50	14	180

TABLE 4. SUGGESTED RELIABILITY LEVELS FOR LIFELINES

LIFELINE		RELIABILITY WITHIN INTENSITY AREAS	
		HIGH INTENSITY GROUND MOTION — (MMI IX-X WITH SURFACE FAULTING)	MODERATE INTENSITY GROUND MOTION — (MMI VI-VII)
Water Supply System	Storage Reservoirs *	No failure that will endanger lives	Fully functional
	For fire fighting	Level A - Adequate storage available	Fully functional
	Treatment facilities	Level A	Fully functional
	Distribution	Level B - Tank trucks available for potable water	Fully functional
Sewage System	Collection *	Level C	Level A
	Treatment facilities*	Level B	Fully functional
Electrical Power System	Generating Plants (Except Nuclear) *	Level A	Fully functional
	Transmission *	Level A	Fully functional
	Terminals	Level B	Fully functional
	Distribution *	Level B	Level A
Natural Gas System	Transmission *	Level A	Fully functional
	Storage (above grade) *	Level A	Fully functional
	Distribution	Level B	Level A
Highway, Freeway & Railroad	Bridges *	Level A - no collapses	Fully functional
	Roadways	Level B	Fully functional
Airports	Control facilities	Level A	Fully functional
	Runways and Taxiways *	Level A	Fully functional
Port Facilities		Level B	Level A
Radio and TV		Level A	Fully functional
Telephone		Level A	Fully functional

NOTES:

Reliability level A: 5% or less of intensity area without service for one day; fully restored in 1 week.

Reliability level B: 20% or less of intensity area without service for 1 week; fully restored in 1 month.

Reliability level C: 50% or less of intensity area without service for 1 week; 20% or less without service for 1 month; and fully restored in 3 months.

Asterisk * : Hazard of loss of life and private property.

SYSTEM VULNERABILITY

The area of seismic risk has developed considerably since the early studies on high-rise building performance [63]. As a result of the San Fernando earthquake and the relatively poor performance of lifelines in that earthquake, seismic risk concepts have been extended to address many different types of utility and transportation systems. Early studies by Panoussis [47] and Taleb-Agha [58, 60] have been supplemented by more recent studies to provide a wealth of procedures and methods for analyzing the seismic risk to lifelines. Many of the available procedures have come from related areas of systems engineering [11, 12], many have been developed for specific lifeline systems [23, 31, 32, 55].

A number of factors will dictate the type of approach considered. In the following subsections, the applications of simulation techniques as well as analytical methods, are discussed. The use of different analysis models for different system objectives is discussed. Examples of system vulnerability studies are also provided at the end of this section.

Methodologies

In general, there are two basic approaches for analyzing the earthquake vulnerability or risk to a lifeline system. The first approach is based on arriving at an analytical solution of system risk. Individual component models are incorporated within the framework of a system model, and system performances are computed based on individual component performances. The solutions of this type of approach are obtained by convolving individual component failure probabilities through a network equation or algorithm. Depending upon the objective of the analysis, the final system probability can represent the probability of different parts of the system being connected, or the probability that different flow units are reached in different parts of the system.

The second approach is based on simulation techniques. The final result is often of the same form as the previous. However, the method used to arrive at this result is quite different. In contrast to the previous approach, this approach attempts to arrive at a solution of system risk by aggregating the effects of a large sample of simulated system <u>states</u>. In a given simulation run, each component of the network is determined to have failed or remained operational through the drawing of a random number and comparison of it to the component's failure probability. A <u>state</u> of the network is thereby defined as the resulting system configuration.

From a computational standpoint, the analytical approach is more desirable. The number of computations required is often considerably less than that needed using the simulation approach. The analytical approach has one shortcoming, however, and that is a lack of model flexibility. Usually, simplifications are needed in the seismic hazard, component vulnerability, and system models to make the analytical approach tractable. Incorporation of complex component models, e.g., hydraulic flow models, would be very difficult using this approach. On the other hand, the simulation approach can handle complex component modeling, in addition to incorporating many types and levels of uncertainty. However, the price paid for this flexibility is the need for larger computer capability.

Relevant Analyses

Generally, four types of network analyses have been developed: graphic, connectivity, operational, and economic.

Graphic Analyses. Graphic representations of lifeline systems can aid both in simplifying some problems and in detecting certain types of problems. The simplest systems may be linear or serial, as indicated in Figure 4a. In such systems, flows may be unidirectional, as in gravity-flow water supply systems, or bidirectional, as in communications systems. In some cases, bidirectionality may be simultaneous, as with telephone systems or full duplex systems where transmission and reception can occur at the same time. In Figure 4a, nodes A, B, and C, and links AB and AC must all be functional for flows to move from A to C, or from C to A. Flow directions may also be reversible (and so add redundancy). In gas and water systems, for instance, normal flow directions may be reversed under emergency response conditions.

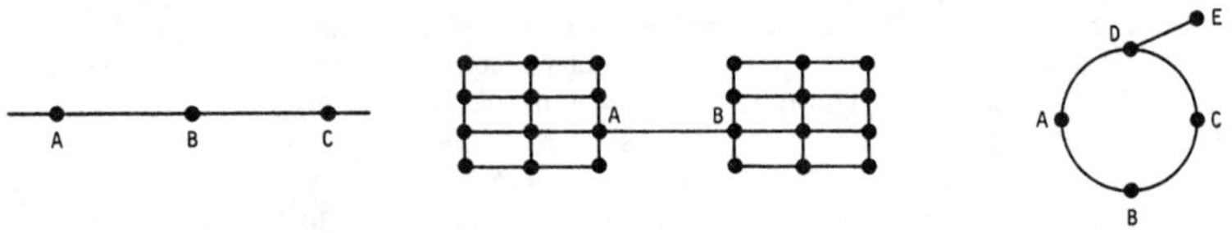

(a) Linear or serial system (b) Linear subsystem (c) Circumferential system

FIGURE 4. GRAPHIC REPRESENTATIONS OF LIFELINE SYSTEM

It is rare, though, for an entire lifeline system to be linear. Oil pipelines, railways, and subways may be exceptions to the usual redundancy in lifeline systems. Still, numerous linear subsystems can be identified within lifeline systems. Figure 4b illustrates one sort of linearity. Here, two grids are connected by link AB, so that their connection can be severed if node A, node B, or link AB are nonfunctioning. The figure may represent two cities connected by a subway, two water systems, and so on.

Even more common are what may be called "circumferential lines," or lines leading away from nodes within a grid or loop, as in Figure 4c. Such a system may represent transmission lines connected to canyon areas or other isolated service areas. Even more common, almost all service connections are so represented.

Connectivity/Operational Analyses. A more elaborate and typically algebraic approach is the connectivity approach, which analyzes networks based on the fact that nodes and links are either connected or not (closed or open). Such an approach was used extensively in the infancy of seismic risk analysis of lifeline networks [47]. The complexities of lifeline grids have led to detailed suggestions for simplification, both for lengthy links and saving computer storage time, [e.g., 58]. Results of connectivity approaches were early held to be applicable to all networks, owing no doubt to their algebraic generality [30, 58].

At present, uses of connectivity approaches appear to be more circumscribed. The needed assumption that nodes are either connected or not may be simplistic for some networks, such as in communications systems wherein long distance high frequency reception may be intermittent or marginal. Even when such an assumption is made, results of connectivity analysis may only be preliminary to further systematic analyses based on flow patterns and emergency operations relative to the unique features of the system.

The need to augment connectivity analyses soon became apparent in regard to electric power systems, and has recently become more feasible in regard to water supply systems. Connectivity, it has been discovered, indicates only that part of a system is nonfunctional, not that the remainder is operational. Systems may be limited to carrying only their maximum load. So, when a portion of the system is inoperative, the remainder may not be able to handle the additional load. In regard to water systems, connection between a supply node and a demand node after an earthquake is compatible with water pressure and flow rates that have dropped below minimum standards.

These considerations place limits on connectivity studies, which nonetheless may be very useful as steps in more comprehensive analyses. Theoretically, at least more comprehensive analyses would incorporate system-specific flow analyses and operational reconfiguration alternatives. Such incorporated elements could in turn reveal existing network deficiencies. Initial attempts at such more comprehensive analyses have been made for both electric power and water supply systems [24, 39, 50, 51, 52, 53, 55].

Economic Analyses. Economic analyses have not yet been made subsequent to full-scale network analyses. However, such analyses involve, first, identification of and economic quantification of systemic benefits of a proposed mitigation; second, a comparison of expected benefits with costs of the mitigation. Examples of systemic benefits may include decreased revenue losses to the utility, decreased secondary losses from systemic failure (fire losses reduced, outage time decreased, and its effects diminished such as through fewer business interruptions), and decreased need for customer to purchase standby equipment for continued utility use. Network analyses turn out to be extremely useful in such economic studies because redundancy in systems often allows continued service in spite of local failures, but in capacitated systems one or a few failures may turn out to have systemic and hence very costly repercussions. To assess the proposed mitigation(s), the network must be analyzed for its expected losses both with and without the proposed improvement.

Output

Three examples are provided in this subsection which illustrate the types of output obtained from a lifeline risk (or reliability) analysis. Each example emphasizes the importance of one of the three major analyses presented in the previous section. (Graphical methods have been excluded here.)

Figure 5 provides an example where connectivity analysis methods were used in analyzing the reliability of a water distribution system [39]. Reliability was defined as the probability of having a connection between a source and a set of demand nodes, or the probability that a particular demand node will receive a minimum level of flow. In order to test the method, several earthquake events and reliability levels were introduced. In addition, to test the usefulness of the method for design and upgrading purposes, the effect of introducing several mitigations was analyzed. The results, which are presented in Figure 5, show that the prescribed mitigations (which are reflected by the strengthened network values) add little to the reliability of the original network. Furthermore, the greatest single threat to lifeline reliability is seen to come from the Hayward fault. Results from this type of analysis can therefore be used to assess the relative effects from different earthquake events, and also to test the sensitivity of various mitigation measures. This type of procedure can also be used to develop the system risk curve presented in Figure 1. Generally, the risk curve is computed as 1 minus the reliability.

Figure 6 provides an example where connectivity and operational methods were used in assessing the earthquake performance of a water supply system [24]. In addition to identifying those nodes that are physically disconnected from a supply source, the analysis also identifies those nodes that are hydraulically sensitive, i.e., cannot be reached because of negative pressure problems. Factors that contribute to these negative pressures include loss of key water supply facilities and pipelines, large service demands, and surviving system configurations. The results of the type of analysis can be used to assess anticipated system performance from specified earthquake events.

Examples of output from an economic analysis are found in Bresko et al. [14], where expected annual total seismic losses to a water system are $0.88 per capita, or 1.5% of system replacement cost. These estimates are then recalculated based on alternative seismicity estimates. Direct losses, such as costs of replacing and repairing damaged pipes, are added to indirect costs of loss of water service, to obtain total costs. Various seismic mitigation programs, such as the gradual replacement of more seismically vulnerable pipes at a given annual rate, could thus be compared to these estimates based on the existing system. The range of possible outputs from such network studies is potentially very great. For instance, if traffic must be rerouted owing to damaged freeway structures, losses to various adjacent businesses may occur, along with other losses from increased fuel costs, driving time, and so on.

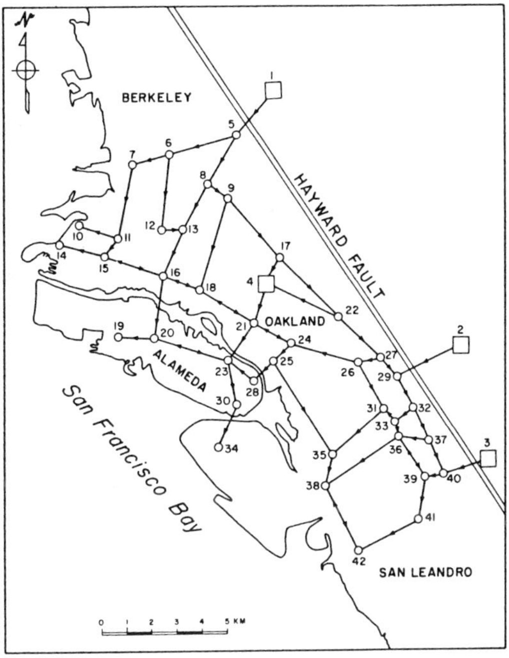

(a) Water distribution system

Case	Connectivity Reliability		Flow Reliability	
	Original Network	Strengthened Network	Original Network	Strengthened Network
Reliability to random earthquake on Hayward fault	0.9867	0.9997	0.9708	0.9913
Reliability to random earthquake on Calaveras fault	1.0000	1.0000	1.0000	1.0000
Reliability to random earthquake on San Andreas fault	0.9989	0.9989	0.9969	0.9970
Reliability to random earthquake on any fault	0.9981	0.9991	0.9954	0.9969
10-year reliability	0.9630	0.9820	0.9111	0.9398
50-year reliability	0.8281	0.9130	0.6277	0.7332

(b) Network reliability results

FIGURE 5. AN EXAMPLE OF A CONNECTIVITY ANALYSIS [39]

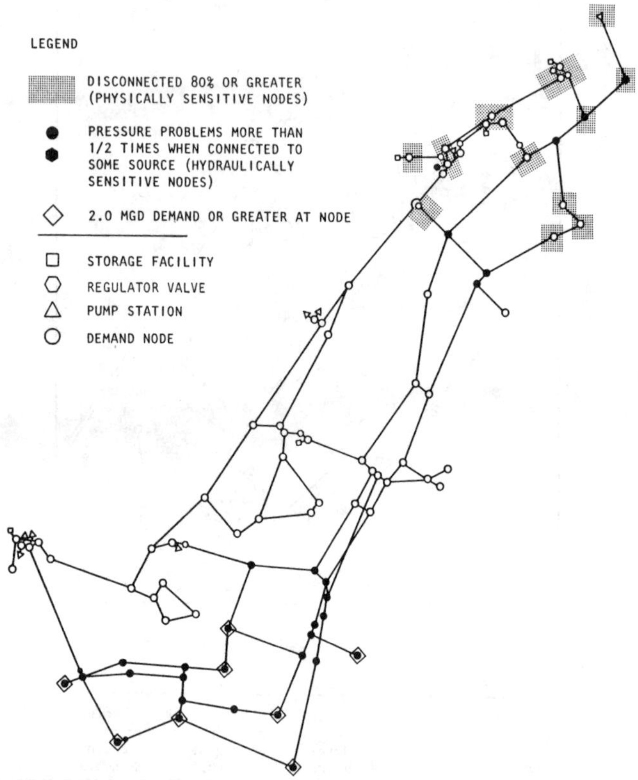

FIGURE 6. NODES AT WHICH EITHER PRESSURE PROBLEMS OR DISCONNECTEDNESS TYPICALLY OCCURRED FOR EARTHQUAKE SITUATIONS [24]

DECISION ANALYSIS

Although not widely applied, decision analysis techniques offer a convenient means of systematically upgrading or designing lifeline systems to earthquake. In the introduction, a general discussion was provided whereby the optimum design level for a lifeline system could be calculated from expected damage and design cost functions. For the most part, this type of analysis has not been performed for lifeline systems because of the lack of such models. Only recently has the state of the art begun to reflect development in this area. With the availability of such models, it should not be long before decision analysis methods are applied in analyzing the earthquake problem.

Design Strategies

Very little work has been done to develop mathematical methods to find the optimum seismic design level for a component or a group of components of a lifeline system. One reason for this, besides the one indicated earlier, is that lifeline components are scattered in large geographical areas and, therefore, a unified risk based design strategy is hard to define. In one study the idea of balancing expected damage with increased design costs was incorporated; this example was a hypothetical water supply system in Southern California [34].

More work has been done to develop methods of finding the optimum seismic design level for discrete components [57, 63]. These methods consider the costs of providing increased seismic resistance, the damage that may occur during future earthquakes and the human and social consequences of such damage. In all of these methods, an optimality criterion is selected and the corresponding optimum strategy is found. Three optimality criteria have been considered. The first is based on a cost-benefit approach where all expected earthquake induced losses are assigned dollar values and the optimum strategy is the one with the minimum total cost. In the second criterion, namely minimum safety, the optimum design strategy is such that the expected loss of lives and injuries is limited to a given value. The third criterion is based on multiattribute decision theory that evaluates alternatives in terms of several characteristics, which means examining the trade-offs between net discounted expected costs of material losses and human lives lost [18, 63, 64]. However, these earlier studies are applicable to facilities located within the same general area with a single seismic hazard curve. Extension of these ideas and methods to cover lifeline networks or systems is still a matter for future research.

Retrofitting and Upgrading Strategies

Suppose it is desired to upgrade an existing lifeline network so as to improve its reliability. Since lifeline networks are usually redundant, the question then arises as to which of the components should be upgraded. These critical components may be defined in different ways.

A simple method has been developed to identify critical components of a network subjected to a set of seismically active source areas and/or faults [59]. This method is based on identifying those components that will cause network failure for each possible seismic event. Integrating over all possible events and normalizing probabilities, one obtains for each network component the probability that this component will be the cause for the network failure. Generally, the component with the highest chance of being the cause of network failure is most critical. Based on this information, an upgrading strategy can be defined. One such strategy would be to upgrade each component with a probability exceeding a certain limit (e.g., 25%, 50%). The extent of upgrading or strengthening of a component could be determined iteratively. That is, after each component is strengthened, a new risk analysis could be performed to determine the new reliability and a new set of critical components. This would be repeated until the desired reliability is reached.

Moghtaderizadeh and Der Kiureghian define a component as critical if a small variation in its strength changes the reliability of the network [40]. The procedure for identifying the critical component is iterative and incremental [39]. It is clear that the upgrading methods mentioned above are based on the idea of providing minimum safety or reliability. Upgrading strategies can be based on other ideas like cost-benefit or optimum allocation of given resources. Such strategies should be consistent, at least in principle, with design strategies of new lifeline systems.

ACKNOWLEDGEMENTS

A significant part of this paper was originally prepared for publication by ASCE/TCLEE as part of an advisory note document on lifeline earthquake design. The present author would like to acknowledge the members of the Seismic Risk Committee of TCLEE for their initial contributions and comments to the original paper. The author would also like to acknowledge the contributions of Dr. Craig Taylor to this paper.

REFERENCES

1. Agbabian Associates, "Procedures and Criteria for Increasing the Earthquake Resistance Level of Electrical Substations and Special Installations," R-7211-2859, El Segundo, CA, 1973.

2. Algermissen, S.T., and Perkins, D.M., "A Probabilistic Estimate of Maximum Acceleration in Rock in the Contiguous United States," Open File Report 76-416, U.S. Geologic Survey, 1976.

3. American Society of Civil Engineers, The Current State of Knowledge of Lifeline Earthquake Engineering, ASCE, New York, 1977.

4. American Society of Civil Engineers, Social and Economic Impact of Earthquakes on Utility Lifelines, ASCE, New York, 1980.

5. American Society of Civil Engineers, Lifeline Earthquake Engineering: The Current State of Knowledge, ASCE, New York, 1981.

6. American Society of Civil Engineers, "Advisory Notes on Lifeline Earthquake Engineering," Technical Council on Lifeline Earthquake Engineers, New York, 1983.

7. American Society of Civil Engineers, "Guidelines for Seismic Design of Oil and Gas Pieplines," Technical Council on Lifeline Earthquake Engineering (Gas and Liquid Fuels Committee), New York, 1984.

8. American Society of Mechanical Engineers, Lifeline Earthquake Engineering - Buried Pipelines, Seismic Risk, and Instrumentation, ASME, New York, 1979.

9. American Society of Mechanical Engineers, Proc. International Symposium on Lifeline Earthquake Engineering, ASME, New York, Jun 1983.

10. Anderson, T.L., and Bachman, R.E., "LNG Terminal Design for California," presented at ASCE National Convention, session no. 46, Seismic Design of Lifeline Structures, Portland, Apr 1980.

11. Barlow, R.E., and Prochan, The Statistical Theory of Reliability and Life Testing, Holt, Rinehart and Winston, New York, 1975.

12. Barlow, R.E., Der Kiureghian, A., and Satyanarayana, A., "New Methodologies for Analyzing Pipeline and Other Lifeline Networks Relative to Seismic Risk," ORC 80-5, Operation Research Center, Univ. of California, Berkeley, Mar 1980.

13. Blume, J.A., et al., "Earthquake Damage Prediction: A Technological Assessment," Report No. 17, Dept. of Civil Engineering, Stanford Univ., 1975.

14. Bresko, D., Hendrickson, C., and Oppenheim, I., "Seismic Risk Analysis of an Urban Water System," Lifeline Earthquake Engineering, Proc. Second Specialty Conf., ASCE, Oakland, CA, Aug 1981.

15. Campbell, K.W., Eguchi, R.T., and Duke, C.M., "Reliability in Lifeline Earthquake Engineering," Journal of the Technical Councils of ASCE, TC2, 1979, pp. 259-270.

16. Culver, C.G., et al., "Natural Hazards Evaluation of Existing Buildings," Building Science Series 61, National Bureau of Standards, U.S. Dept. of Commerce, 1975.

17. Czarnecki, R.M., "Earthquake Damage to Tall Buildings," Report No. R73-8, Dept. of Civil Engineering, Massachusetts Institute of Technology, 1973.

18. De Neufville, R., "How Should We Establish Public Policy on Setting Design Codes for Performance Standards?" Proc. U.S. National Conf. on Earthquake Engineering, Ann Arbor, MI, 1975, pp. 327-336.

19. Duke, C.M., and Matthiesen, R.B., "Earthquakes, Lifelines and ASCE," Civil Engineering, ASCE, 1973, pp. 63-67.

20. Duke, C.M., and Moran, D.F., "Earthquakes and City Lifelines," San Fernando Earthquake of February 9, 1971 and Public Policy, Special Subcommittee of the Joint Comm. on Seismic Safety - Calif. Legislature, 1972.

21. Duke, C.M., and Moran, D.F., "Guidelines for Evolution of Lifeline Earthquake Engineering," Proc. U.S. National Conf. on Earthquake Engineering, Ann Arbor, MI, 1975, pp. 367-375.

22. Eguchi, R.T., et al., "Earthquake Vulnerability of Water Supply Systems," Lifeline Earthquake Engineering: The Current State of Knowledge, ASCE, New York, 1981, pp. 277-292.

23. Eguchi, R.T., "Seismic Vulnerability Models for Underground Pipes," Proc. International Symposium on Lifeline Earthquake Engineering, ASME, Portland, Jun 1983.

24. Eguchi, R.T., et al., "Earthquake Performance of Water Supply Systems," Proc. International Symposium on Lifeline Earthquake Engineering, ASME, Portland, Jun 1983.

25. Environmental Quality Systems, Inc., "Earthquake Design Criteria for Water Supply and Wastewater Systems," Rockville, MD: EQS, 1980.

26. Erel, B., et al., "Measuring the Earthquake Performance of Urban Water Systems," The Current State of Knowledge of Lifeline Earthquake Engineering, ASCE, New York, 1977, pp. 183-198.

27. Ford, L.R., and Fulkerson, D.R., Flows in Networks, Princeton Univ. Press, 1962.

28. Hall, W.J., and Newmark, N.M., "Seismic Design Criteria for Pipelines and Facilities," Journal of the Technical Councils of ASCE, Nov 1978, pp. 91-107.

29. Hasselman, T.K., Eguchi, R.T., and Wiggins, J.H., "Assessment of Damageability for Existing Buildings in a Natural Hazards Environment," Report No. 80-1332-1, J.H. Wiggins Company, Redondo Beach, CA, 1980.

30. Hein, K.H., and Whitman, R.V., "Effects of Earthquakes on System Performance of Water Lifelines," R76-23, SDDA Report No. 27, Dept. of Civil Engineering, Massachusetts Institute of Technology, 1976.

31. Hendrickson, C., Kufert, D., and Oppenheim, I., "Water System Network Analysis Under Seismic Hazard," Reprint 80-C2/PUP-62, ASME, New York, 1980.

32. Hendrickson, C., Oppenheim, I., and Siddharthan, K., "User Losses in Earthquake-Damaged Roadway Networks," Journal of the Technical Councils of ASCE, Aug 1980.

33. Isoyama, R., and Katayama, T., "Practical Performance Evaluation of Water Supply System During Seismic Disaster," Lifeline Earthquake Engineering: The Current State of Knowledge, ASCE, New York, 1981, pp. 111-126.

34. Jacobsen, S.E., Torabi, M., and Bansal, P.P. "Earthquake Damage Costs in the Design of Water Resource Systems," Water Resources Research, Vol. 10, No. 2, Apr 1974.

35. Kachadoorian, R., "Earthquake: Correlation Between Pipeline Damage and Geologic Environment," Journal of the American Waterworks Association, 1979, pp. 165-168.

36. Katayama, T., Kubo, K., and Sato, N., "Earthquake Damage to Water and Gas Distribution Systems," Proc. U.S. National Conf. on Earthquake Engineering, Ann Arbor, MI, 1975, pp. 396-405.

37. Kennedy, R.P., Darrow, A.C., and Short, S.A., "Seismic Design of Oil Pipeline Systems," Journal of the Technical Councils of ASCE, 1979, pp. 119-134.

38. Longinow, A., Bergmann, E., and Cooper, J.D., "Bridge Retrofitting: Selection of Critical Bridges in a Road Network," The Current State of Knowledge of Lifeline Earthquake Engineering, ASCE, New York, 1977, pp. 83-96.

39. Moghtaderizadeh, M., et al., "Seismic Reliability of Lifeline Networks," Journal of the Technical Councils of ASCE, ASCE, Vol. 108, No. TCI, 1982, pp. 60-78.

40. Moghtaderizadeh, M., and Der Kiureghian, A., "Reliability Upgrading of Lifeline Networks for Post-Earthquake Serviceability," Proc. Seventh European Conf. on Earthquake Engineering, Athens, Greece, Sep 1982.

41. McCaffrey, M., and O'Rourke, T.D., "Buried Pipeline Response to Reverse Faulting During the 1971 San Fernando Earthquake," Proc. International Symposium on Lifeline Earthquake Engineering, ASME, Portland, Jun 1983.

42. National Oceanic and Atmospheric Administration/Earthquake Engineering Research Institute Earthquake Investigation Committee, "Earthquake Damage to Water and Sewage Facilities," San Fernando, California, Earthquake of February 9, 1971, Vol. II, NOAA, 1973, pp. 73-193.

43. Newmark, N.M., and Hall, W.J., "Pipeline Design to Resist Large Fault Displacement," Proc. U.S. National Conference on Earthquake Engineering, Ann Arbor, MI, 1975, pp. 416-425.

44. Noda, S., et al., "A Decomposition Method for Lifeline Risk Analysis," Lifeline Earthquake Engineering: The Current State of Knowledge, ASCE, New York, 1981, pp. 171-186.

45. O'Rourke, M., Singh, S., and Pikul, P., "Seismic Behavior of Buried Pipelines," Lifeline Earthquake Engineering - Buried Pipelines, Seismic Risk, and Instrumentation, ASME, New York, 1979.

46. O'Rourke, T.D., and Trautmann, C.H., "Analytical Modeling of Buried Pipeline Response to Permanent Earthquake Displacements," School of Civil and Environmental Engineering, Cornell University, 1980.

47. Panoussis, G., "Seismic Reliability of Lifeline Networks," Report No. 15, Dept. of Civil Engineering, Massachusetts Institute of Technology, 1974.

48. Robinson, R.R., Longinow, A., and Chu, K.H., "Seismic Retrofit Measures for Highway Bridges," FHWA-75-79-216, Dept. of Transportation, Washington DC, 1979.

49. Salvadori, M.G., and Singhal, A., "Strength Characteristics of Jointed Water Pipelines," Interim Grant Report No. IR-3, Weidlinger Associates, Menlo Park, CA, Jul 1977.

50. Schiff, A.J., and Yao, J.T.P., "Response of Power Systems to the San Fernando Valley Earthquake of 9 February 1971," Center for Large-Scale Systems, Purdue Univ., Jan 1972.

51. Schiff, A.J., "Earthquake Effects on Electric Power Systems," Journal of the Power Division, ASCE, Vol. 99, No. PO2, Nov 1973, pp. 317-328.

52. Schiff, A.J., Feil, P.J., and Newsom, D., "Computer Simulation of Lifeline Response to Earthquake," Proc. Sixth World Conf. on Earthquake Engineering, New Delhi, India, 1977, Vol. III, pp. 3298-3302.

53. Schiff, A.J., "Seismic Emergency Operation of Power Systems," Journal of the Technical Councils of ASCE, Nov 1978, pp. 1-11.

54. Shinozuka, M., and Koike, T., "Estimation of Structural Strains in Underground Lifeline Pipes," Lifeline Earthquake Engineering - Buried Pipelines, Seismic Risk, and Intrumentation, ASME, New York, 1979.

55. Shinozuka, M., Tan, R., and Koike, T., "Serviceability of Water Transmission Systems Under Seismic Risk," Lifeline Earthquake Engineering: The Current State of Knowledge, ASCE, New York, 1981.

56. Taylor, C.E., Eguchi, R.T., and Wiggins, J.H., "Lifeline Earthquake Engineering: State-of-the-Art of Hazard Mitigation Analysis," Proc. Third International Earthquake Microzonation Conf., Seattle, Vol. 111, 1982.

57. Taleb-Agha, G., "Sensitivity Analyses and Graphical Method for Preliminary Solution," Res. Rept. R74-42, SDDA Report No. 14, Dept. of Civil Engineering, Massachusetts Institute of Technology, 1974.

58. Taleb-Agha, G., "Seismic Risk Analysis of Networks," Res. Rept. R75-43, SDDA Report No. 2, Dept. of Civil Engineering, Massachusetts Institute of Technology, 1975.

59. Taleb-Agha, G., "Seismic Risk Analysis of Lifeline Networks," SDDA Report No. 24, Dept. of Civil Engineering, Massachusetts Institute of Technology, 1975.

60. Taleb-Agha, G., "Seismic Risk Analysis of Lifeline Networks," Bulletin of Seismological Society of America, Vol. 67, No. 6, Dec 1977, pp. 1625-1645.

61. Task Committee on Research Needs of the Committee on Research of TCLEE, "Research Needs in Lifeline Earthquake Engineering," Journal of the Technical Councils of ASCE, Dec 1979, pp. 343 ff.

62. Wang, L.R.L., and Cornell, H.A., "Effect of Earthquakes on Buried Pipelines," Proc. American Waterworks Association Annual Conference and Exposition, San Francisco, 1979.

63. Whitman, R.V., et al., "Seismic Design Decision Analysis: Methodology and Pilot Application," Res. Rept. R74-15, Dept. of Civil Engineering, Massachusetts Institute of Technology, 1974.

64. Whitman, R.V., and Cornell, C.A., Chapter 9 - "Design," Seismic Risk and Engineering Decisions, ed. C. Lomnitz and E. Rosenblueth, Elsevier Scientific Pub. Co., New York, 1976, pp. 339-380.

65. Whitman, R.V., and Hein, K.H., "Damage Probability for a Water Distribution System," The Current State of Knowledge of Lifeline Earthquake Engineering, ASCE, New York, 1977, pp. 410-423.

Retrofitting Lifeline Facilities: The Japanese Approach

A.J. Schiff[1], M, P.I. Yanev[2], M, and H.H. Givin[3].

ABSTRACT

A brief history of the prediction of a major earthquake in the Tokai Area of Japan is reviewed as is the governmental response to the prediction. The holistic approach to planning and mitigation incorporating all levels of government and the private sector is a key feature of the Japanese approach. Emphasis is placed on mitigation measures used to improve the earthquake response of lifeline facilities. The lifelines discussed include power, transportation, and gas pipelines systems.

Introduction

This paper is based on efforts of the Committee on the Tokai Earthquake (1), Earthquake Engineering Research Institute (EERI), which was formed to investigate the extensive planning and mitigation effort mounted by the Japanese in response to the expectation of a major earthquake in the Tokai region of Japan. The history of the prediction effort and the governmental response to it is briefly reviewed. The holistic approach to planning and mitigation incorporating all levels of government and the private sector is a distinctive feature of the Japanese approach to disaster planning. Emphasis is placed on mitigation measures used to improve the earthquake response of lifeline facilities. The lifelines discussed here include power, transportation, and gas pipelines systems. In addition, modifications which are common to many lifeline systems will be reviewed. A more complete and detailed report discussing the findings of the entire committee will be issued by EERI (2).

1 Prof., School of Mechanical Engineering, Purdue University, West Lafayette, Indiana

2 President, EQE Inc., San Francisco, California

3 IBM, General Products Division, San Jose, California

Scientific Foundation for the Earthquake Prediction

The area along the southern coast of Japan centered about 100 miles west of Tokyo, referred to as the Tokai region, has been subjected to a series of major earthquakes associated with a subduction zone between the Philippine Sea Plate and the Eurasian Plate. Several sequences of earthquakes along the fault zone have occurred, each sequence separated by about a hundred years. At the present time an earthquake is expected in the Tokai region on an unruptured fault segment, or seismic gap, laying between the 1944 Tonankai and the 1946 Nankaido earthquake fault zones. Long term observations of deformations in the region indicate that there is significant accumulated strain associated with this seismic gap. Also, a low rate of occurrence of small and moderate magnitude earthquakes in the offshore Tokai region is similar to that in the adjacent seismic gap prior to the 1944 Tonankai earthquake. It is anticipated that the earthquake will occur in the next 10 years and that it will have a Richter Magnitude of about 8. It is expected that the earthquake will occur in the Suruga Trough.

The Tokai Region

The Tokai region forms part of the main transportation corridor between Tokyo to Osaka and is a modern industrial area. There is a high population concentration along the coast. Shizuoka City is the largest city in the region. The Tokai area encompasses about 14,400 square km, with about six million inhabitants in 170 cities, towns, and villages. In addition to its role as a vital transportation link between Tokyo and Osaka, numerous large corporations have important manufacturing facilities in the region.

The area designated for Intensified Measures is defined as the area of greatest damage resulting from the Magnitude 8 Ansei earthquake of 1854. Figure 1 which has been taken from (2), shows the area of Intensified Measures, its proximity to the suspect fault and the distribution of the various governmental units contained in the area. It is an area where the damage is anticipated to be equal or exceed an Intensity of about IX on the Modified Mercalli Scale.

Governmental Disaster Response

In 1961 the Disaster and Countermeasures Basic Law was passed which was directed at mitigating all natural catastrophes, including earthquakes. Under the provisions of this law the Central Disaster Prevention Council, consisting of the Prime Minister and 18 agency heads, was formed and given broad emergency responsibilities and powers. For example, the national government can mandate specific actions and expenditures to be made by prefecture (county) and municipal governments.

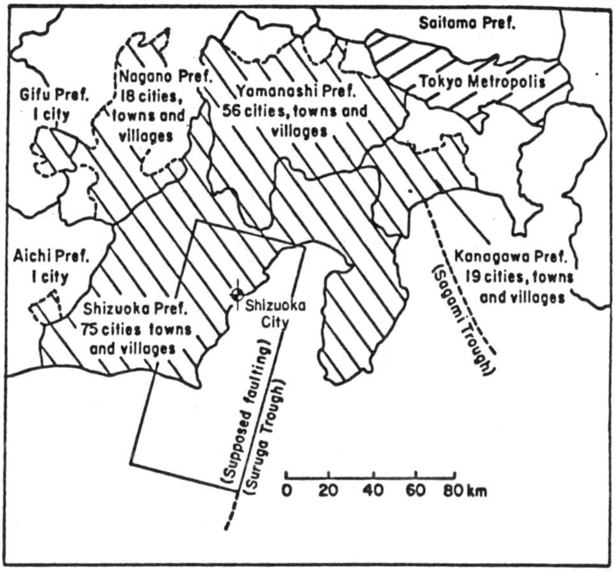

The Tokai Region

The observation in 1974 that the probability of a repetition of the 1854 Ansei earthquake in the Suruga Trough was increasing and the earthquake prediction made by the Chinese in 1975 stimulated the formation of a Special Committee on Earthquake Disaster Prevention Measures. The committee recommended legislation to strengthen earthquake preparedness programs in 1977. An Earthquake Assessment Committee consisting of six seismologists was formed in 1977 for predicting the Tokai earthquake. In 1978 the Large-Scale (or Intensive) Earthquake Countermeasures Act was introduced and passed by the national legislature.

The large-Scale Earthquake Countermeasures Act calls for several distinctive programs for areas designated for "Intensified Measures." These include increased scientific investigations directed at earthquake prediction and a range of preparedness activities to mitigate the effects of the earthquake. The national government has the lead responsibility for administering preparedness planning, retrofitting of significant public buildings, and development of tsunami defenses. This is done in association with prefectural and municipal governments, which are mandated by the Act to participate. It requires development of "short-term" plans by all levels of government and certain private industry to increase public safety.

The total earthquake counter measures budget for all of Japan is approximately $3 billion annually, broken down as indicated below.

Prediction research	$25,751,000
Training	12,075,000
Improvement of Facilities	2,525,107,200

It should be emphasized that budget information is often not directly comparable with U.S. budget figures. For example, these figures do not include salaries. Also, funds spent by the private sector for improved earthquake response are not included.

In August 1979, the Tokai area was designated for intensified measures under the act. This has triggered a massive program. Governmental funds for the Tokai program for the 1979-1984 total $1,716,738,100. These funds are for the prediction and preparedness effort including a long 6 meter tsunami wall in Shizuoka, retrofitting public buildings, installing a wireless communications network, and educational activities. While the national government supports many of the activities, a significant burden falls on the prefecture and local governments. To help support this, Shizuoka Prefecture established a 10 percent increase in the surtax on corporate projects for a five year period starting in 1979. The proceeds of this tax, about $4 million a year, is split between the Prefecture and local governments. It does not include scientific research for prediction or salaries of officials involved in the preparedness effort.

Prediction: Methods and Procedures

A key assumption to much of the preparedness effort is that there will be a short-term (1 to 3 days) warning. To this end, instrumentation from on-shore and permanent ocean bottom seismometers in to Tokai area are monitored by the Japan Meteorological Agency (JMA), the agency delegated this responsibility under the Countermeasures Act. In addition, information from volumetric down-hole strain meters, tiltmeters, tide gauges, and ground water wells are used. This data is continuously monitored. If criteria that have been established are exceeded, the Earthquake Assessment Committee (EAC) is summoned and meets within 30 minutes. Within 30 minutes after the EAC has convened, the Japan Broadcasting Corp. will issue an announcement by radio and television throughout the Tokai area that the EAC has just been called into emergency session.

The chairman of the EAC has the responsibility of making a recommendation to the director of the JMA. The director of JMA will advise the Prime Minister, who will be standing by with his cabinet. A unanimous decision of the cabinet is required before a warning statement can be issued. It is estimated that this entire prediction process, culminating in the Prime Minister's statement, will take less than two hours.

The issuance of an official earthquake warning by the Prime Minister sets into action a complex chain of events. The special wireless communications system disseminates the warning. As a result, trains in the area proceed to the nearest safe station and stop. The telephone communications system changes to a more reliable mode of operation so as to assure the availability of service to high priority phones. Within the Tokai area almost all social and economic functions would be suspended except for electricity, gas and water services. Public transportation would be stopped, reduced speed limits would be imposed and roads are to be free of traffic within 2 hours. Schools will be evacuated and closed. Areas of high fire and tsunami risk would be evacuated. Provisions have been made for sustaining individuals in staging areas for three days.

Public participation has also been extensive. This participation has not only included education programs but the mobilization of a large part of the population into voluntary community groups. For example, in Shizuoka, 96% of the residents are resistered in one of over 4000 groups. These groups are involved in conveying the warning and are supplied with support equipment such as portable pumps, fire extinguishers, etc.

There has been extensive retrofitting to structures with emphasis on schools, government buildings, and emergency facilities. While in some cases buildings are rebuilt, typically shear walls are added, foundations strengthened and heavy roofing replaced with lighter weight materials. Retrofitting has run about 10% of the replacement value of the reworked structures.

Industry has also taken many steps to improve the earthquake response of its facilities and to protect its workers. For example, all companies in Shizuoka Prefecture, about 17,000, have been required to submit emergency response plans. While these are only given a cursory review due to limited governmental resources, it forced each company to review its own situation. Exemplary plans for several companies have been published as a book to provide guidance to other companies. The vast majority of the cost of corporate preparedness and retrofitting is born by the individual companies. Some of these actions will be discussed under lifeline measures. It is clear that larger companies that have the resources have done extensive modifications. For example, one manufacturing company has secured all machine tools to their foundations, marked and protected evacuation routes within the plants, strengthened storage tanks for hazardous materials, etc.

Lifelines

Lifelines in Japan not only perform the same function as in the U.S., but their configuration, facilities, and equipment typically are similar to that in the U.S. As in the U.S., lifelines are owned both by the public and privates sectors, depending on the particular lifeline. For example, communications and electric power are privately owned.

Scope of Earthquake Preparedness Effort

One of the most notable features of the Japanese approach to earthquake mitigation under the Large-Scale Earthquake Countermeasures Act is the broad scope of the effort. Organizations that are brought into the planning process include the national government, local governments, corporate management, the community and the family. A broad range of activities are also involved included planning, earthquake prediction, public education, and mitigation.

The remainder of this section will briefly summarize some of the activities which characterize the Japanese approach to disasters. The following section will deal with prediction and warning in more detail.

Both national and prefectural governments within the area of Intensified Measures are directly involved in counter measures planning through programs which includes strengthening public buildings, assisting in damage assessments and risk analysis, allocating financial support to establishing local disaster supply centers, developing and coordinating preparedness and prediction response plans, establishing and operating a wireless communication network for rapid dissemination of the prediction/warning announcement, developing public education campaigns, assisting private industry in developing their own plans, and carring on basic research in the natural and social sciences.

A vital element which significantly shaped the character of the program is the assumption that a short-term earthquake prediction will be issued. Key to much of the effort is the response to the earthquake prediction. A wireless system which makes extensive use of facsimile and standard messages to enhance the reliability of the communications has been installed. Pre-earthquake warning announcements will be given, however, it is only after the Prime Minister officially declares the warning can the various prediction response plans developed by governments and industry be put into effect. Vital to the effectiveness of any emergency plan is the need for it to be exercised. This is done, to a degree, twice a year, once during a National Earthquake Day and a second time during local exercises.

One part of the preparedness effort is the assessment of earthquake hazards so that appropriate planning and mitigation actions can be taken. A devastating tsunami accompanied the 1854 earthquake. On the basis of the run up associated with this earthquake, a 6 meter tsunami wall is being constructed along much of the coast in the Shizuoka Prefecture. Extensive evacuation procedures have also been developed for the communities that are in the tsunami inundation zone. Also, because of the high building density and type of construction, Japan has a severe fire hazard. Again, evacuation routes from hazardous areas have been identified and hardened against fire. Where required, new staging areas have been created by acquiring and clearing land. Maps indicating other hazards, such as landslides and soil liquefaction have been prepared and made available.

The importance of lifelines is universally accepted. In Japan the importance of the knowledge of their response in the post-earthquake environment for planning purposes is also appreciated so that the Shizuoka Prefectural government has estimated and published disruption times for the various lifelines. This is in marked contrast to the U.S. where it is occasionally suggested, often implied, and generally assumed by different groups that there will be minimum disruption of lifeline services.

Published estimates of disruption times may provide a rude awakening to individuals concerned with the planning in the U.S., particularly in light of the fact that the state of planning and earthquake resistance of Japanese facilities is generally above that in the U.S. It is anticipated that power will be 1/3 restored after 5 days and 2/3 restored after 10 days. Individuals are informed that they should have stored sufficient water for three days. Water for vital uses will be restored in 4 to 7 days and other water will be temporarily restored in the period from 8 to 30 days. Important and public telephones will be restored after 7 days with most service restored after 30 days. It is recognized that if the gas system is shut down, it may require up to a month for the restoration of normal service, even if there is no earthquake damage.

Power Systems

While power utilization patterns, power network layout, and strategies for configuring power grids in Japan and the U.S. differ, equipment types, function, form, and construction materials closely parallel each other. In addition, substations, the system elements which has proved to be most vulnerable to earthquake damage, are layed out in similar ways in Japan and the U.S.

The attitude of the electric power industry towards earthquakes in Japan is markedly different from that in the U.S. Factors that contribute to this are the fact that a large percentage of the population is located in regions of moderate to high seismic risk, higher seismicity increases the potential for damage, and earthquake damage has been experience more frequently. This heightened concern about earthquakes can be illustrated by several examples. The Central Research Institute of the Electric Power Industry (CRIEPI) in Japan, which is the counterpart to the Electric Power Research Institute (EPRI) in the U.S., has had an active research program directed at mitigating earthquake effects. This activity has extended beyond the problems confronting the nuclear industry. Japanese companies have initiated significant internal research programs directed at the seismic hazard. Major power equipment manufacturers have significant research programs directed at improving the earthquake resistance of their products.

Chubu Electric Company, the utility that services the Tokai area, has initiated extensive modifications to substation equipment as a result of the Tokai earthquake prediction. For example, air blast circuit breakers and air switches are being retrofitted by adding three diagonal stays between the top of tall ceramic members and their bottom support. To do this, the equipment is returned to the manufacturer for modification. Also, transformer bushings have be redesigned to improve their strength and old units are being replaced by redesigned members. Thin rubber pads, originally placed under transformers to reduce noise, have been removed. It has been estimated that equipment retrofitting costs about 10% of the replacement cost of the equipment. Very little retrofitting has been done to fossil fuel power plants other than bracing control panels.

The Hamaoka nuclear power plant is located in the area of the highest expected accelerations form the Tokai earthquake, which are in excess of 0.4g. The first two units went into commercial operation in 1976 and 1978. They are both boiling water reactors of the General Electric Mark 1 type with ratings of 540 MWe and 840 MWe. Unit three, rated at 1100 MWe, is under construction. Units 1 and 2 were designed for peak ground acceleration at zero period of 0.3g and 0.45g for the two earthquake levels used for the design of nuclear plants in Japan. The corresponding criteria for Unit 3 were increased to 0.45g and 0.6g. It is planned that the reactors will be shut down if a prediction is issued for the Tokai earthquake. The conventional power plants in the Chubu Electric Company will remain on line to supply power to the system.

The nuclear plants have extensive strong motion instrumentation which should provide valuable information when the earthquake occurs. Units 1 and 2 have 70 data channels and Unit 3 has about 200 channels, including a down hole array below the plant.

Gas Systems

In part, because of the experience Japan has had with post-earthquake fires, the post-earthquake operation of gas systems is markedly different from that in the U.S. Shizuoka Gas Company has installed peak acceleration detectors at various locations in its system. If a peak acceleration between 150 gals and 200 gals is detected, the system is put on alert. If a peak acceleration between 200 gals and 250 gals is detected, the pressure in gas transmission lines is reduced and LPG storage tanks are isolated. If the peak acceleration is between 250 gals and 300 gals, gas generation is stopped and valves to the transmission system are closed so that the system is sectionalized. The situation is also reviewed to determine if the gas transmission pipe lines are to be purged. If the peak acceleration exceeds 300 gals, the transmission lines will be purged. It is estimated that if this is done, it will take a month to get the system back to normal, even if there is no system damage. About 430 liquid propane gas (LPG) tanks are located in the area of intensified measures. It is considered that these tanks represent a substantial hazard and an extensive strengthening program has been undertaken. A review indicated that 260 of the tanks required strengthening. Special criteria for strengthening have been developed and all strengthening is expected to be completed by 1986. Much of the work is quite expensive because many of the tanks are located in areas that are expected to liquefy and improved foundations are required.

Transportation Systems

Major transportation facilities exist in this region, which is part of the main transportation corridor between Tokyo and Osaka. The well-known Shinkansen, or Bullet Train, which runs along the coast in this area, and is paralleled by the Tomei Expressway, the major limited-access highway between Tokyo and Osaka. As a result of the Tokai earthquake prediction, significant retrofitting is occurring to transportation systems, particularly bridges. For bridges within Shizuoka Prefecture, about 75% of the 840 have been identified as needing upgrading, usually due to expected liquefaction. About a third of these retrofits will be completed by 1985. For multi-span bridges, the ends of alternate spans are being fixed to their support structures and large slotted links connect the unanchored ends of adjacent spans. The tops of piers have been enlarged, so that spans should not fall off the pier even if displacements exceed those allowed for by the sliding joints. Framed-bent piers have had shear walls retrofitted to the bents, or have had their bearing area significantly increased. In some cases, additional piles have been driven, to mitigate expected liquefaction effects.

Conclusions

The purpose of this paper, which emphasizes lifelines, is to inform the U.S. earthquake community of the Tokai earthquake prediction and the mitigation measures associated with it. In reference to lifeline, the Japanese effort has demonstrated their strong commitment to planning and preparedness and the willingness of the private sector to make significant expenditures to strengthen equipment and facilities.

References

1. Members of the EERI Tokai Committee are V.V. Bertero (Chairman), C. Scawthorn (Secretary), R. Andrews, C. Arnold, H.H. Givin, A.J. Schiff, P. Somerville, P.I. Yanev, R.E. Scholl.

2. "A Report of the Future Tokai Earthquake," Committee on the Tokai Earthquake, EERI, to be published.

Acknowledgements

The gracious hospitality and sharing of information of the officials of all of the organizations that we visited while in Japan is greatly appreciated. Funds were provided by the National Science Foundation for the committee to visit Japan and the assistance of Messrs. T. Owens and M Miyahara of the Tokyo office was most helpful.

T. Iwasaki of the UJNR and his staff at the Public Works Research Institute provided liaison in Japan and guided our visit to Shizuoka. NBS and the US UJNR Panel on Wind and Seismic Effects, especially H.S. Lew and R.D. Marshall, provided coordination for the Committee's visit. FEMA is also thanked for its assistance.

CSMIP STRONG-MOTION INSTRUMENTATION AND RECORDS FROM TRANSPORTATION STRUCTURES - BRIDGES

A.F. Shakal[1], J.T. Ragsdale[2] and R.W. Sherburne[1]

ABSTRACT

Seven transportation structures have been instrumented by the California Strong Motion Instrumentation Program (CSMIP), including four bridges. The instrumentation configurations are presented for these bridges, with a discussion of the planning guidelines used in selecting and instrumenting the structures. Significant records (in excess of 0.25 g maximum acceleration) have been obtained from three highway bridges. At one bridge, a freeway overpass on Hwy 101 in Rio Dell (near Eureka in northern California), strong shaking has been recorded from three separate earthquakes. Preliminary analyses of a subset of these records indicate that the vertical response of the bridge superstructure was much greater than the horizontal response.

INTRODUCTION

The damage that occurred to highway bridges during the San Fernando earthquake of February 9, 1971 demonstrated that the response of these structures during earthquake shaking was poorly understood. The behaviour of these bridges led to the initiation of an effort within the California Strong Motion Instrumentation Program (CSMIP) to instrument bridges and other transportation structures in the mid-1970s. Several goals were considered in planning the instrumentation in place today. Early analyses of bridge response data by Raggett and Rojahn (6) and the general planning concepts introduced by Rojahn and Raggett (8) were major contributors to present instrumentation concepts. There are two principal factors in the planning of strong motion instrumentation - the selection of the facility to be instrumented, and the location of sensors on the facility.

SELECTION AND INSTRUMENTATION OF STRUCTURES

Selection of Structure

The selection of a structure to be instrumented is typically made on the basis of the type of the structure and its location. In the selection of buildings for instrumentation, a basic assumption has been that the most desirable strong motion records would be those obtained

---

1) Seismologist, Office of Strong Motion Studies, Calif. Div. Mines and Geology, 2811 O Street, Sacramento, CA 95816.

2) Structural Engineer, Office of State Architect, Structural Safety Section, 1500 Fifth Street, Sacramento, CA 95814.

from structures which undergo damage from an earthquake. This approach leads directly to implications on location: the most desirable locations are those near enough to a fault of sufficient potential that severe shaking may be expected at the site in the lifetime of the structure.

The type of structures selected for instrumentation depends on at least two factors. The structure should be simple, and it should be typical in design. The value of a simple structure is that fewer assumptions are needed in interpreting or modelling the recovered data. A structure which is typical of many others is of value because the lessons learned from the interpretation of data from one structure are transferable to other similar structures.

Locating Sensors on a Structure

Rojahn and Raggett (8) list three approaches which can be used to guide the planning of instrumentation for a selected structure. These include: force level determination, mathematical model identification, and mathematical model verification. Each of these approaches require differing amounts of instrumentation and provide differing levels of detail about the motion of the structure. The first, force-level determination, requires an extensive amount of instrumentation if accurate force levels are to be determined from the recorded data without resort to mathematical models of the structure. The second, model identification, also requires extensive instrumentation because the objective is to actually develop a mathematical model of the structure from the recorded data. The last approach, model verification, requires the least instrumentation because only selected measurements are desired to verify particular aspects of an existing mathematical model for the structure.

Transportation structures instrumented by CSMIP have mostly been planned with either force-level determination or model verification in mind. Perhaps after detailed analysis of the response of several transportation structures the more limited model-verification instrumentation approach will see wider application.

INSTRUMENTED TRANSPORTATION STRUCTURES

Seven transportation structures have been instrumented in basic accordance with the considerations and guidelines discussed above. These seven structures, listed in Table 1, include a tunnel, a wharf and an airport tower in addtion to four bridges. The tunnel, Caldecott Tunnel in the east-Oakland area, is a part of a major surface traffic artery. The instrumentation of this structure was completed in 1979 and is described by Ragsdale and McJunkin (7), and in general by Brekke (1). Surface and downhole (275 feet) triaxial accelerometers measure input earthquake ground motion at a site 390 feet northwest of the tunnel center. Thirteen sensors are located within the tunnel to record its response. They were mounted at three locations within the tunnel: three sensors were located near the east portal, seven were located 1,040 feet in from the east portal, and three were located about 1,600 feet in from the east portal, near the center of the tunnel.

TABLE 1

TRANSPORTATION FACILITIES INSTRUMENTED BY CSMIP

Structure	Description	N.Lat. W.Long.	No. Chns	Date Installed	No.Records, Max.Accel.
San Juan Bautista Overpass	Hwy 101/156 Separation Bridge	36.86N 121.58W	12	5/77	1, 0.33g
Rio Dell Overpass	Hwy 101/Painter St Overpass	40.50N 124.10W	20	9/77	3, 0.66g
Meloland Overpass	Hwy Interstate 8 Overpass	32.77N 115.45W	26	4/78	1, 0.50g
Caldecott Tunnel	Hwy 24 Tunnel, east of Oakland	37.86N 122.21W	19	12/79	2, 0.03g
Vincent Thomas Bridge	Suspension bridge, Long Beach	33.75N 118.27W	26	10/81	0
Oakland Wharf	14th St. Wharf, Port of Oakland	37.82N 122.31W	12	1/84	0
Lancaster Air Control Tower	Fox Field Tower, Lancaster	37.74N 118.21W	9	3/84	0

The Oakland Wharf is a long (1,620 feet) concrete, pile-supported, deck structure 65 feet wide. Two triaxial sensors are located on the ground, approximately forty feet from the deck to measure input motion. Six horizontal sensors have been placed along the deck to measure its transverse and longitudinal response.

The Lancaster Airport Tower is a sixty foot high braced steel structure typical of the design used by the Federal Aviation Admin. (FAA) at many small airports in California. A triaxial sensor package was placed at the base of the structure to measure the input at ground level. Horizontal and torsional response of the structure is measured by three sensors below the control cab and three sensors on the roof of the cab.

Future plans for CSMIP include the instrumentation of 40 transportation structures of various types in California. Bridges are the most common type of transportation structure instrumented to date. The bridges instrumented by CSMIP are described in the following sections, in addition to the strong motion data recorded to date.

INSTRUMENTED BRIDGES: OBJECTIVES AND SENSOR CONFIGURATIONS

Table 1 indicates that a total of four bridges have been instrumented by CSMIP in the last seven years. The most recent is also the most complex - the Vincent Thomas Bridge, a three-span suspension bridge in Long Beach. The schematic illustration of the location of the 26 sensors on the structure is shown in Figure 1. This bridge is the only suspension bridge instrumented by CSMIP. The instrumentation was a cooperative effort between the CSMIP and the USGS. A report by C. Rojahn is in preparation on this project, which is not further discussed here.

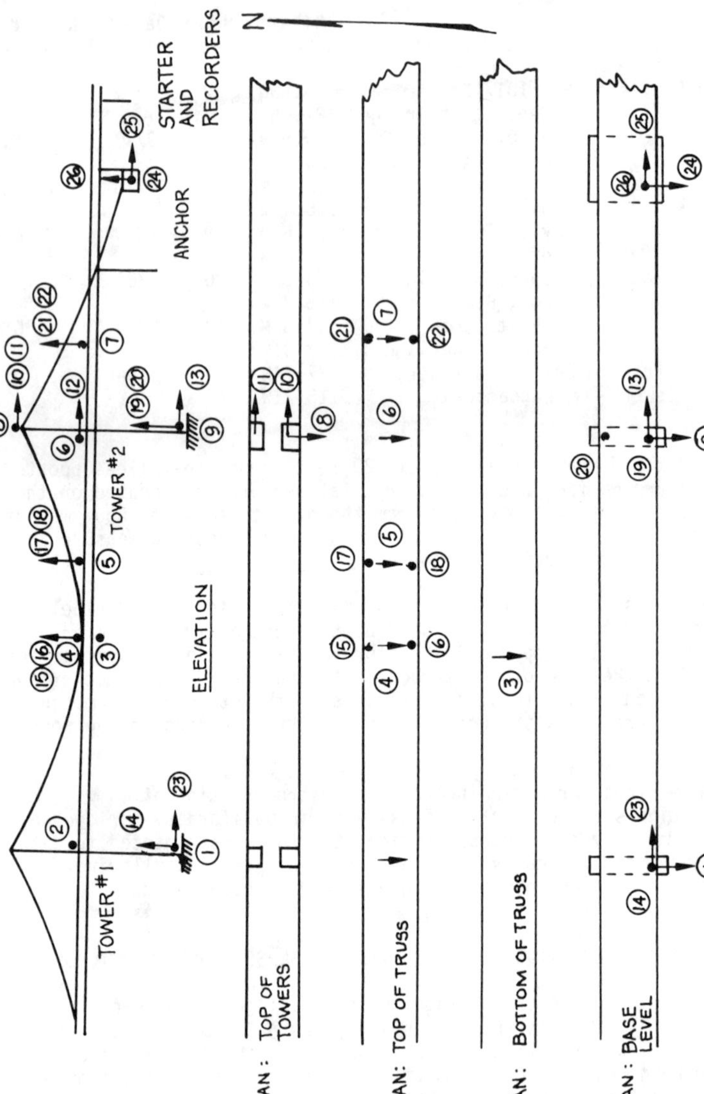

Figure 1. Schematic of strong-motion instrumentation on the Vincent Thomas Bridge in Long Beach. The locations and orientations of the 26 accelerometers are indicated by arrows and dots, which represent arrows normal to the plane of the figure.

Three freeway overpasses are presently instrumented by the CSMIP. The Meloland Overpass, shown in Figure 2, is a good example of a simple freeway overpass. The bridge is a continuous two-span cast-in-place reinforced-concrete box girder with spans approximately 105 feet in length. Figure 2 shows a total of 26 sensors (force-balance accelerometers); these are recorded on two 13-channel analog film recorders. This heavily instrumented structure is a good example for considering in some detail the factors involved in choosing appropriate sensor locations on the structure.

The three principle approaches in strong-motion instrumentation, as discussed above, are force-level determination, model identification, and model verification. Force-level determination and model identification both require relatively extensive instrumentation. For either of these approaches, Rojahn and Ragget (8) note that the number, location and orientation of sensors required on the structure depend on 1) the anticipated modes of structural response; 2) the extent of symmetry in the structure; 3) the number of locations needed to adequately characterize the input motion, and 4) the potential failure mechanisms.

The layout of sensors for the Meloland Overpass in Figure 2 is a good example of the application of these guidelines. The free-field motion is recorded by three sensors (14,15,24) approximately 200 feet from the bridge. Input motion, as well as the response of the embankments, are recorded by the triaxial sensors at the south (10-12) and north (23,25,26) embankments. Input motion is also measured at the base of the central support column (1,2,4). The remainder of the sensors, on the bridge itself, are primarily oriented vertically or horizontally (transverse to the bridge). The verticals along the east edge of the deck (19,16,17,20,6) provide the vertical motion along that edge of the bridge, and coupled with the corresponding sensors on the west edge of the deck (18,21,22) provide for a complete picture of the first-mode vertical motion of the spans and the torsional motion of the deck. The transverse horizontal response of the deck is provided by the transverse sensors along the east edge of the deck (3,5,7,9,13). Finally, the sensors at the foot of the central support column (1,2,4) provide a reference for relative motion of the bridge superstructure relative to the base of the column. Some longitudinal response information is provided by sensors 4 and 25.

This structure was instrumented by CSMIP during 1978, with sensor layout planned by C. Rojahn and J. D. Raggett of the U.S. Geological Survey and J. H. Gates of the Califonia Department of Transportation. Fortunately, during the following year the Imperial Valley earthquake occurred (15 Oct 1979) and resulted in an extensive set of data from the bridge. These data (4) have been used in several research projects (e.g., Lisiecki (3), Rojahn and others (9)).

In contrast with the Meloland Overpass, instrumented well enough for model identification studies, the San Juan Bautista Overpass has relatively limited instrumentation. This bridge, shown in Figure 3, is a long five-span structure, with instrumentation most appropriately described as model verification since only two of the five bents are instrumented. Good records were recovered from this bridge during the

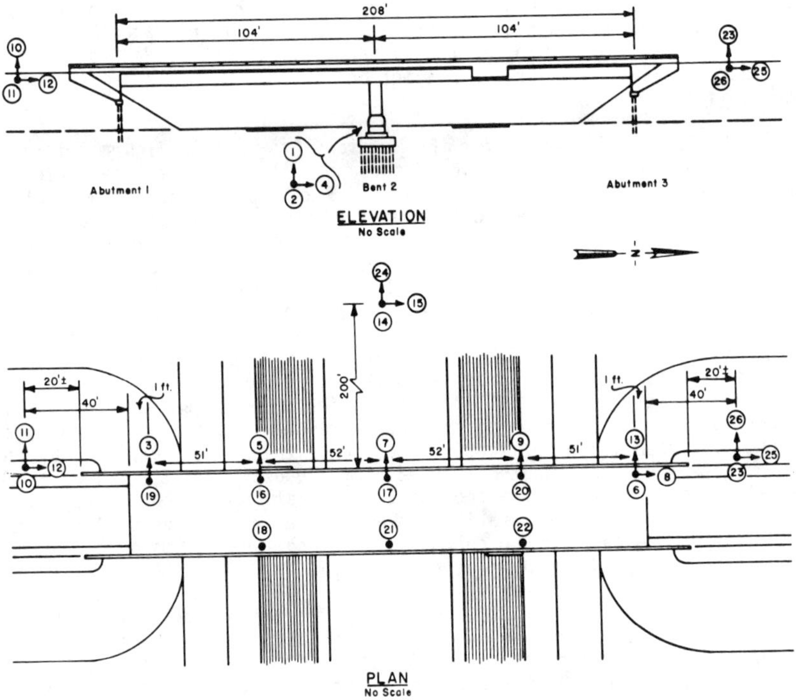

Figure 2. Schematic of strong-motion instrumentation installed on the Meloland Overpass over Hwy 8 near El Centro in southern California. Faulting associated with the 15 Oct 1979 Imperial Valley earthquake passed within 0.5 km of this overpass.

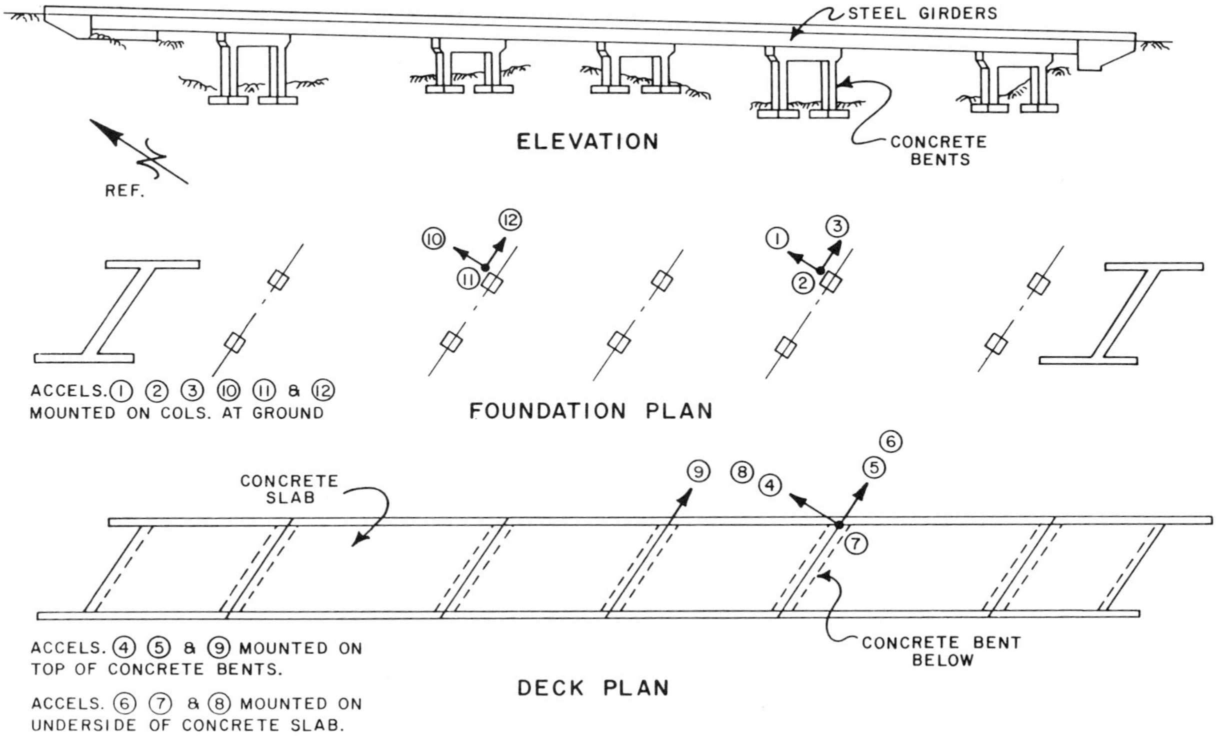

Figure 3. Schematic of strong-motion instrumentation installed on the San Juan Bautista Overpass, a highway 101/156 separation bridge near Gilroy.

1979 Coyote Lake earthquake and have been digitized (Porter and others, (5)) and subsequently analyzed by several investigators.

The third freeway overpass instrumented by CSMIP is located in Rio Dell (near Eureka) on Hwy 101 (see Figure 5). This structure, shown schematically in Figure 4, is relatively simple like the Meloland Overpass, except that the longitudinal axis of the bridge is skewed relative to the highway passing underneath. The instrumentation layout is similar in many respects to that of the Meloland Overpass. One difference is that only one edge of the bridge deck is instrumented, so that twisting or torsion of the deck cannot be measured. At the east end of the bridge triaxial sets of sensors are located both on the embankment and end of the bridge deck so relative motion between the deck and the embankment can be measured.

These three overpass bridges, Meloland, San Juan Bautista and Rio Dell, have each recorded strong shaking since they were instrumented in the late 1970s. The Rio Dell overpass has recorded strong shaking three times since it was instrumented, making it of particular interest.

STRONG MOTION RECORDS FROM THE RIO DELL OVERPASS

The Rio Dell Overpass was shaken by the large (6.9ML) Trinidad-Offshore earthquake of November 8, 1980 and by two additional events since then. The locations and magnitudes of the earthquakes are indicated on the map of Figure 5 and listed in Table 2. The maximum accelerations recorded at the bridge was over 0.25g for each event, as indicated in Table 2. The earthquake magnitudes and distances vary from the local Rio Dell 4.4ML event of 16 Dec 82, at a distance of 15 km, to the 6.9ML Trinidad-Offshore event of 8 Nov 1980 at a distance of 70 km. Records from the most recent event, the 5.5ML Cape Mendocino-Offshore event of 24 Aug 1983, have been digitized recently and are discussed in the following.

TABLE 2

EARTHQUAKES RECORDED BY THE RIO DELL OVERPASS INSTRUMENTATION

Earthquake	Date, Time (GMT)	Location	Depth (km)	Mag. (ML)	Epicentral Distance	Max. Accel.
Trinidad-Offshore	8 Nov 1980, 10:27	41.00N 124.64W	9	6.9	72.	0.34g
Rio Dell	16 Dec 1982, 06:53	40.37N 124.06W	5	4.4	15.	0.60g
Cape Mendocino-Offshore	24 Aug 1983, 13:36	40.31N 124.77W	30	5.5	61.	0.27g

Cape Mendocino-Offshore Earthquake

The instrumentation schematic in Figure 4 indicates that the Rio Dell overpass has a series of vertical sensors along the south edge of the bridge deck, as discussed above. The vertical accelerations recorded at the middle of each span and at the end of the abutments

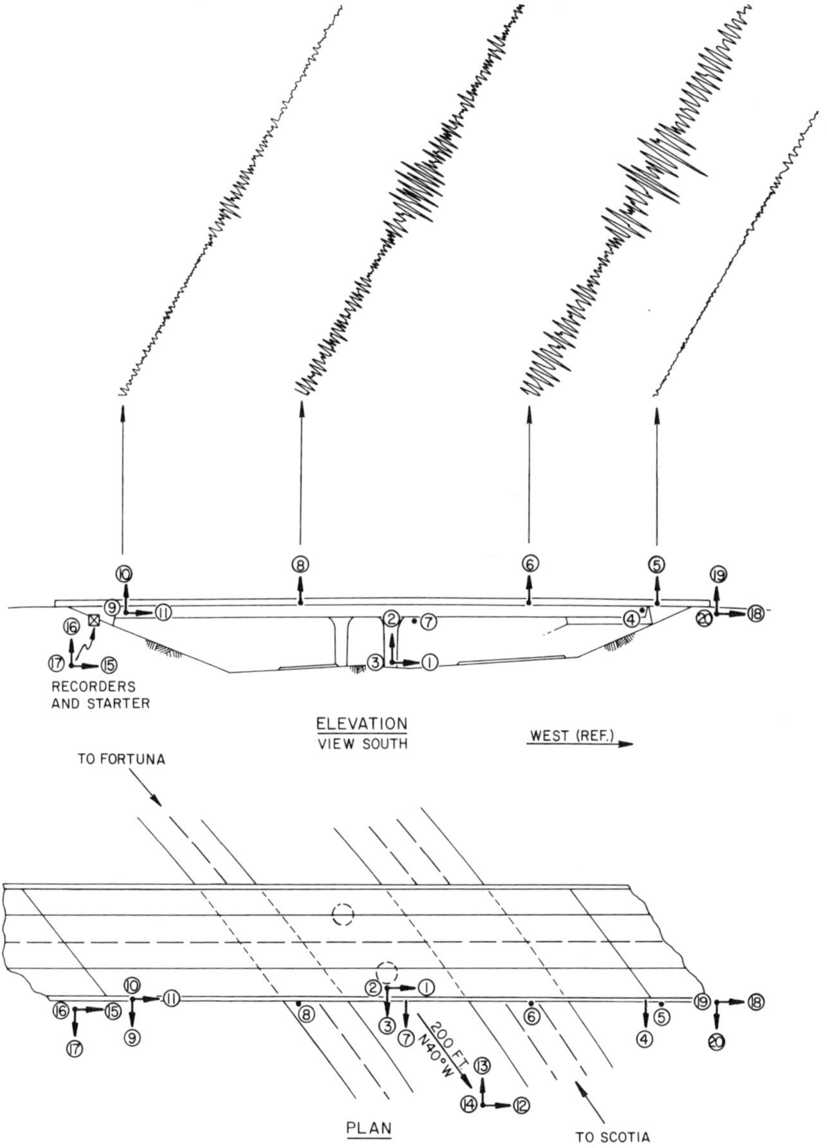

Figure 4. Strong-motion instrumentation installed on the Hwy 101 overpass in Rio Dell (near Eureka in northern California). The upper part of the figure compares the vertical accelerations recorded at the abutments and at the centers of the spans during the 24 Aug 1983 earthquake.

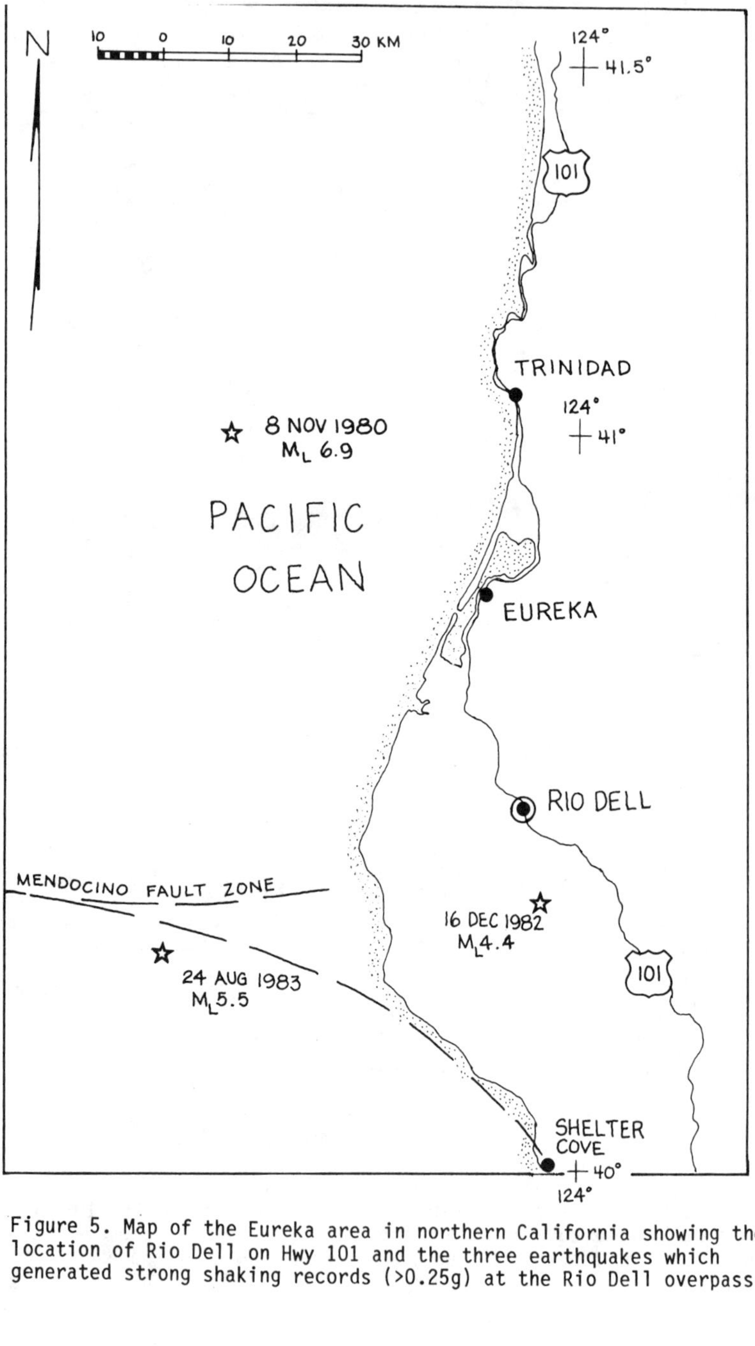

Figure 5. Map of the Eureka area in northern California showing the location of Rio Dell on Hwy 101 and the three earthquakes which generated strong shaking records (>0.25g) at the Rio Dell overpass.

(sensors 5,6,8,10) during the Cape Mendocino-Offshore event are shown in Figure 4. The peak vertical acceleration at the center of the west span (sensor 6) is larger than at the abutment (sensor 5) by nearly a factor of ten. The response spectra for these channels are shown in Figure 6, and they also show a striking difference, with the mid-span spectra higher by a factor of 6 or more at periods shorter than about 0.5 sec (frequencies higher than 2 Hz).

The instrumentation also includes a series of transversely horizontal sensors along the bridge deck. Figure 7 shows the transverse accelerations recorded on the abutments and at the middle of the bridge near the center bent (sensors 4, 7 and 9). In contrast with the dramatic difference seen for the vertical response of the bridge, the transverse accelerations show only a moderate increase in amplitude at the center of the bridge compared to the abutments. The spectra for the west abutment and center bent records are compared in Figure 8. Although the center bent spectra shows some increase over the abutment record at frequencies above 2 Hz, the difference is very much less than for the vertical response shown in Figure 6.

Trinidad-Offshore Earthquake

The largest earthquake recorded in California since the 1952 Kern County earthquake was the 6.9ML Trinidad-Offshore earthquake of 8 Nov 1980. The rarity of large earthquakes make that an important event. The nature of ground motion from large earthquakes is indicated by the displacement record from the free-field instruments of the Rio Dell instrumentation, shown in Figure 9. The long-period nature of the ground motion and its long duration are expected properties of large-earthquake records not present in the many existing records from moderate earthquakes.

That the response of the Rio Dell bridge was recorded during this earthquake makes the records very valuable. This earthquake caused the collapse of two spans of the Fields Landing Overpass on Hwy 101, about 23 km (15 mi) north of Rio Dell. As discussed by Imbsen (2), that overpass failed because of relative displacements between the abutments and deck spans. The Rio Dell instrumentation includes sensors on the embankment and the deck oriented to record the longitudinal motion of the deck relative to the embankment. The longitudinal displacements in the free-field, on the east embankment, and on the east end of the deck (sensors 12,11,15) are shown in Figure 9. It is apparent that there is little difference between the displacements at these locations for this bridge, for this earthquake.

CONCLUSION

Seven transportation structures have been instrumented by CSMIP including three freeway overpasses, a suspension bridge, a tunnel, a wharf and an airpot tower. Strong motion records in excess of 0.25g maximum acceleration have been obtained from the three highway bridges. One of the bridges has been shaken by three separate earthquakes ranging from a large, distance earthquake to a small local event. Preliminary analysis of the records indicates a strong response of the bridge spans in the vertical direction, with only moderate response in the transverse horizontal direction.

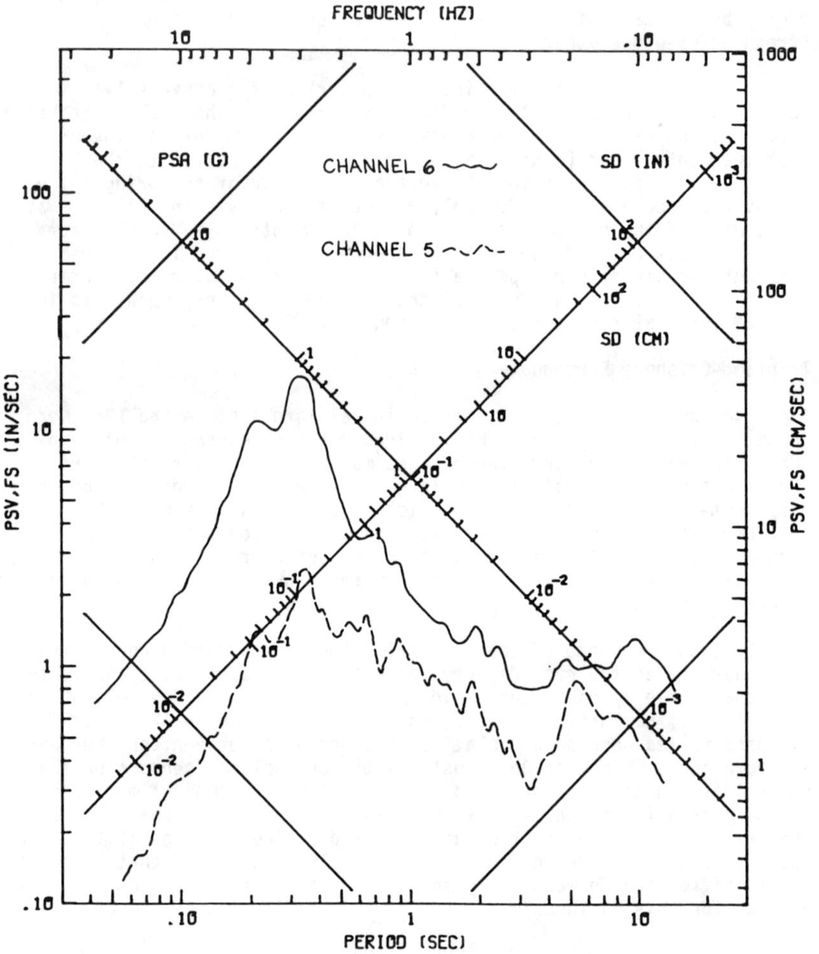

Figure 6. Response spectra (5% damping) for the mid-span (solid, sensor 6) and west abutment record (dashed, sensor 5) of the Rio Dell overpass for the 24 Aug 1983 Cape Mendocino-Offshore earthquake.

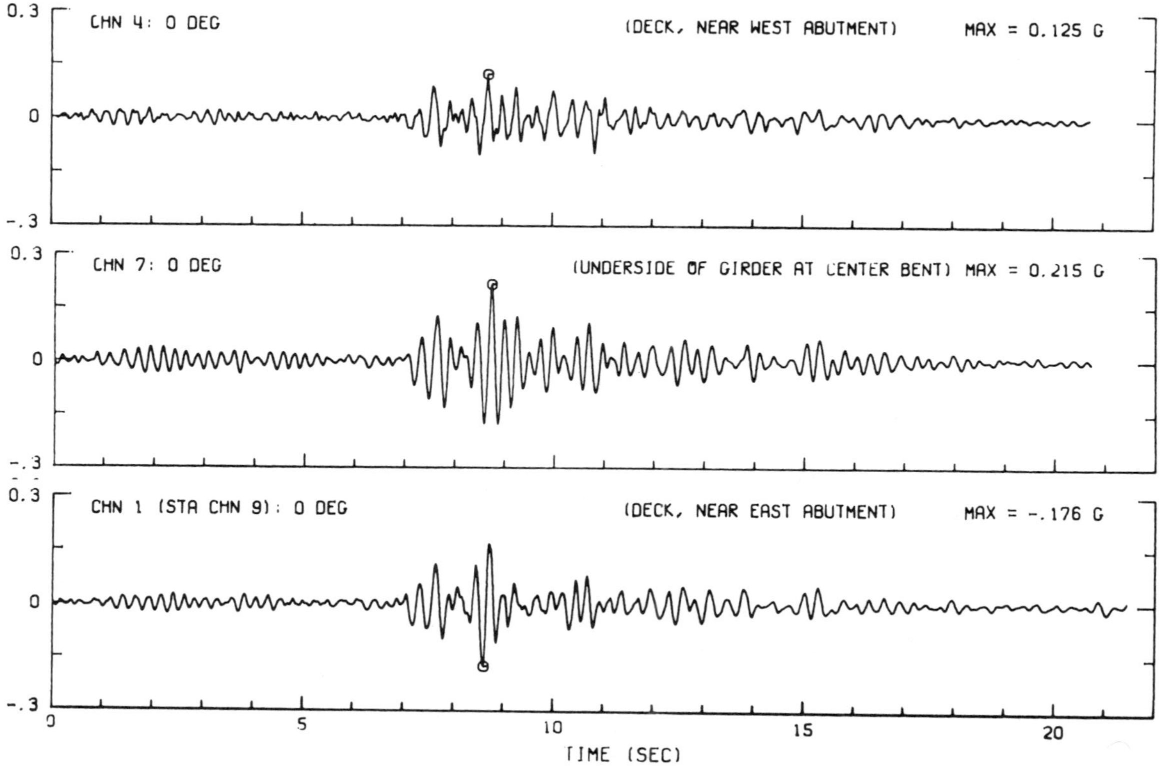

Figure 7. Horizontal transverse acceleration records from the west abutment (upper), bridge center (middle) and east abutment (lower) of the Rio Dell overpass for the 24 Aug 1983 earthquake.

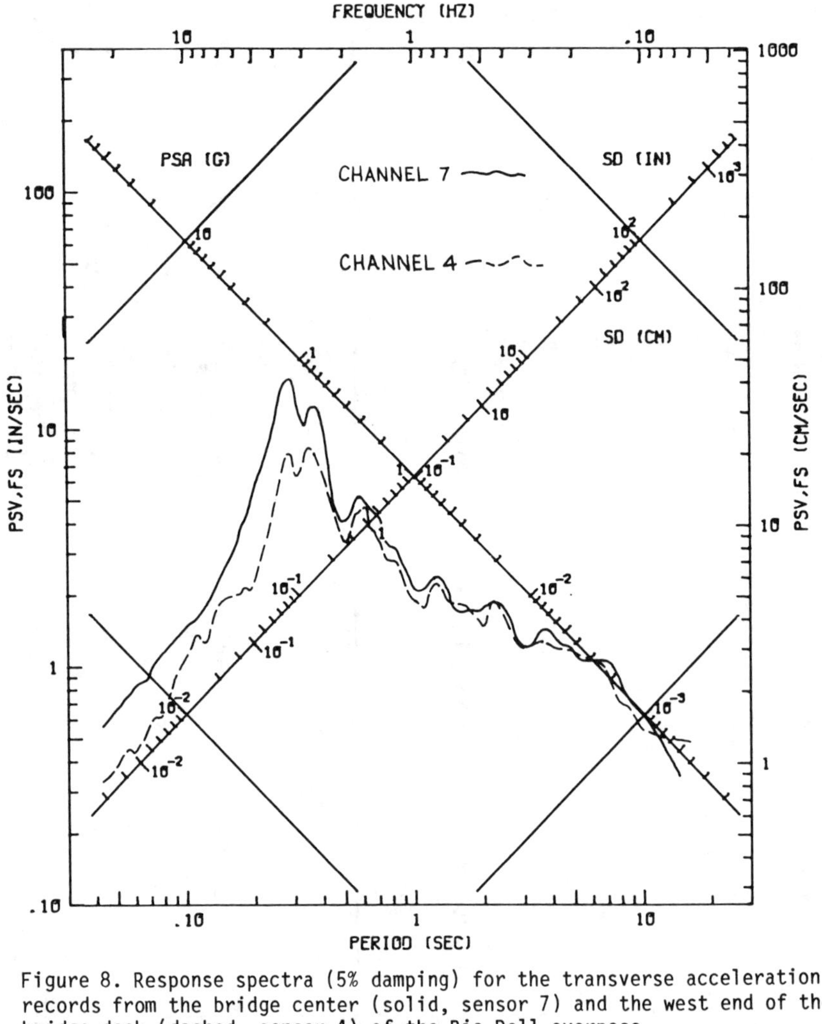

Figure 8. Response spectra (5% damping) for the transverse acceleration records from the bridge center (solid, sensor 7) and the west end of the bridge deck (dashed, sensor 4) of the Rio Dell overpass.

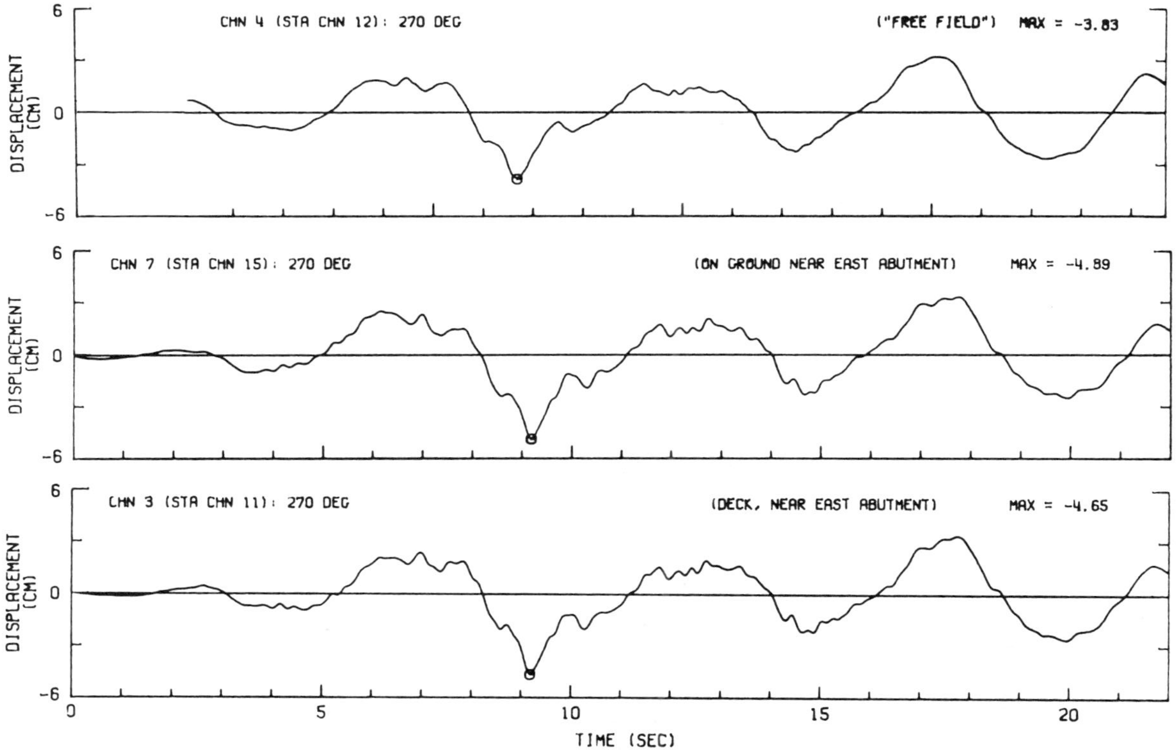

Figure 9. Longitudinal displacements computed from the accelerations recorded at the free-field sensor, at the east embankment, and at the east end of the Rio Dell bridge deck during the 6.9ML Trinidad-Offshore earthquake of 8 Nov 1980.

REFERENCES

1. Brekke, T.L., 1977, Seismic instrumentation of transportation tunnels in California, Federal Highway Administration Report FHWA-RD-77-138, 22 pp.

2. Imbsen, R.A., 1981, Trinidad-offshore, California earthquake, structural damage to bridges, Earthq. Engr. Research Institute Newsltr, vol. 15, no. 1B, p.82-101.

3. Lisiecki, L.C.C, 1982, Analysis of the Meloland overcrossing response to the October 15, 1979 Imperial Valley earthquake, Univ. Calif. Irvine, School of Engnrng, M.S. dissertation, 94 pp.

4. Porter, L.D., 1983, Processed data from the strong-motion records of the Imperial Valley earthquake of 15 October 1979, Calif. Div. Mines Geol., Office of Strong Motion Studies, Special Publication 65, 317 pp.

5. Porter, L.D., A.G. Brady. P.N. Mork and V. Perez, 1983, Processed data from the San Juan Bautista 101/156 separation bridge and the San Juan Bautista freefield records from the Coyote Lake earthquake, 6 August 1979, Calif. Div. Mines Geol., Office of Strong Motion Studies Report SP-64 (with U.S. Geological Survey).

6. Raggett, J.D. and C. Rojahn, 1978, Use and interpretation of strong-motion records from highway bridges, Federal Highway Administration Report FHWA-RD-78-158, 168 pp.

7. Ragsdale, J.T. and R.D. McJunkin, 1981, Seismic instrumentation of Caldecott tunnel, Federal Highway Administration Report FHWA-RD-81, 47 pp.

8. Rojahn, C. and J.D. Raggett, 1981, Guidelines of strong-motion instrumentation of highway bridges, Federal Highway Administration Report FHWA-RD-82-016, 122 pp.

9. Rojahn, C., J.T. Ragsdale, J.D. Raggett and J.D. Gates, 1980, October 15, 1979 main-shock strong-motion records from the Meloland Road-Interstate Route 8 overcrossing, Imperial County, California, U.S. Geological Survey Open-File Report 80-1054, 15 pp.

SEISMIC DESIGN OF AIRPORT CONTROL TOWERS

by

Arthur B. Andersen, P.E.[(1)], Dennis H. Baerwald, P.E.[(2)],
and Bharat R. Soli, P.E.[(3)],

ABSTRACT

The paper describes state-of-the-art design of Airport Control Towers in seismically active areas. It includes a review of current code design parameters, specific tower design criteria and concepts plus several examples of recent tower designs. Tower design examples include Anchorage International Airport, Alaska; Fairchild Air Force Base, Washington; Edwards Air Force Base, California; and Mather Air Force Base, California.

INTRODUCTION

Federally owned towers and support buildings at airports are designed to conform to local, state and/or military building codes. Seismic design considerations, however, significantly impact the structural and architectural design process as applied to air-traffic control towers. Currently there are no comprehensive design criteria for traffic control towers subjected to earthquakes.

A comparison of lateral forces due to wind and earthquake in seismic Zones I through IV, reveals a dramatic increase in the applied load envelopes specifically for Zones III and IV. (The equivalent static force method of design is assumed for determining comparable base shear forces.) A corresponding increase in tower base overturning moments is also evident which mandates special attention on tower foundation design. Allowable tower drift resulting from applied wind and seismic lateral forces is a particularly important design consideration. Drift must be reduced to minimize any movement or vibration that might cause an interruption of communication. This latter criteria often controls selection of tower material and shape. Specific user requirements such as tower occupancy, height, sound transmission limits, low maintenance and fire resistance also severely impact the designers selection process. Site location is also a critical link in the design process with considerations for specific material, equipment and labor availability, site accessibility, local weather extremes, and construction logistics.

(1) Principal, Andersen-Bjornstad-Kane-Jacobs, Inc., Seattle, WA.
(2) Supervising Structural Engineer - Fairchild AFB & Edwards AFB
(3) Supervising Structural Engineer - Anchorage International

DESIGN CRITERIA

Lateral Loads

Figure 1 represents a comparison between lateral wind and seismic forces for a 150 ft. high concrete framed tower with plan dimension of 23 ft. x 23 ft. Lateral wind pressures are based on ANSI A58.1-1982 minimum design loads for exposure C, an 80 mph basic wind speed, and an Importance Factor of I=1.07. Lateral seismic forces are based on an exposed, concrete framed tower with V = ZKCSIW[1]; Where Z = 0.188 (Zone I) to 1.0 (Zone IV). K = 1.0, CS = .14 (T = 0.50 sec.), I = 1.50, W = tower weight (= 1725 kips), V_I = .04W, V_{II} = .08W, V_{III} = .16W, V_{IV} = .21W

In the above equation, the importance factor I for equivalent lateral seismic force should be 1.5 for all designs. In the event of a major seismic disturbance the air traffic control tower must remain operational and stable to maintain facility communications.

The building period "T" should be calculated initially using equation 3-3A of AFM 88-3; $T = 0.05\ h/\sqrt{D}$. After the tower frame and configuration is confirmed a more accurate analysis should be used such as equation 3-3; $T = 2\pi\sqrt{\left(\sum_{i=1}^{n} w_i \delta_i^2\right) \div \left(g \sum_{i=1}^{n} f_i \delta_i\right)}$

By utilizing a computer model of the tower frame the elastic model deflections are readily available for use in the above formula with the intent to verify the designer's lateral load calculations. For the above tower example equation 3-3A indicates a building period of 1.56 seconds and a corresponding base shear value in seismic zone IV of 210 kips. Conversely, equation 3-3 yields a period value of approximately 0.50 seconds and a base shear magnitude of 362 kips. Due to varying tower shapes and the inherent distribution of mass (i.e. unoccupied, narrow shafts versus occupied floors throughout the height of the building acting as diaphragms to increase stiffness of shaft,) the correlation between the two period calculations and resultant base shear can widely vary.

Table 1 below summarizes the tower base shear forces and overturning moments for the concrete framed tower:

Table 1

		Seismic			
	Wind	Zone I	Zone II	Zone III	Zone IV
Base Shear	136^k	69^k	138^k	276^k	362^k
Base overturning moment	$11{,}000^{k-'}$	$7{,}000^{k-'}$	$14{,}000^{k-'}$	$28{,}000^{k-'}$	$37{,}000^{k-'}$

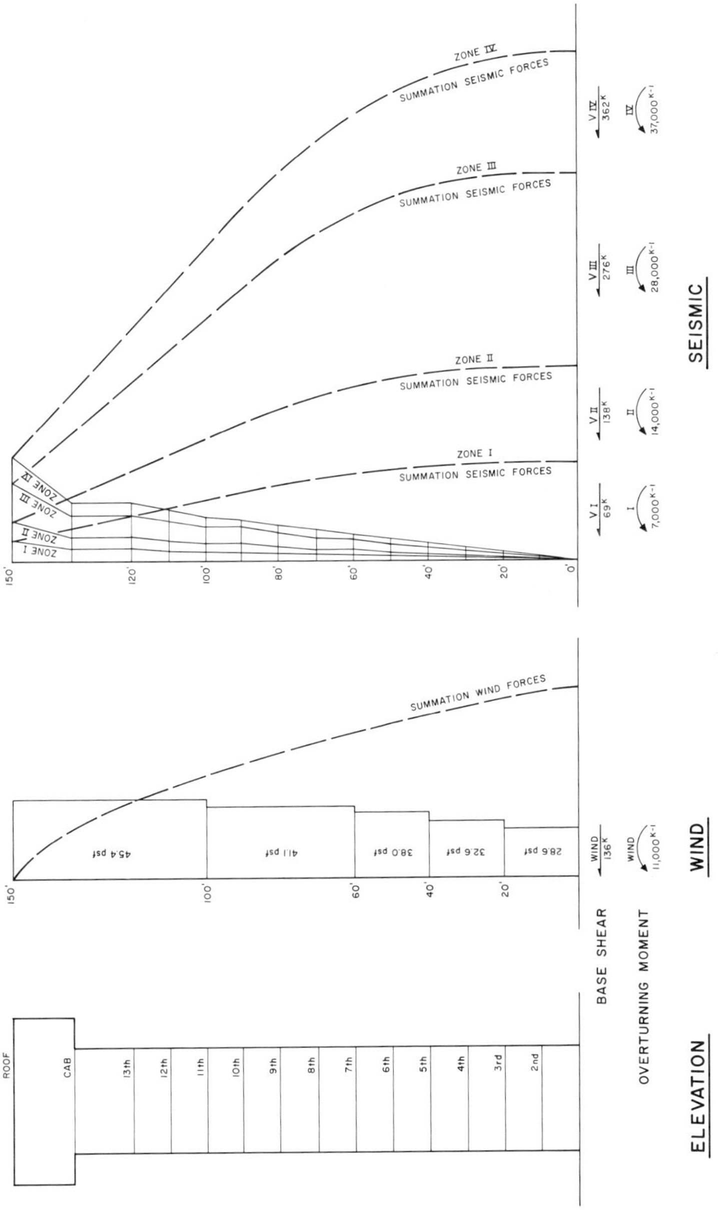

Fig. I. LATERAL FORCES - WIND VS. SEISMIC (CONCRETE TOWER)

Figure 1 is representative of varying tower height designs in that for a high mass concrete framed tower, the seismically induced lateral forces can exceed normal wind forces in seismic Zones III and IV by over 200% and 266% respectively. Due to the greater percentage of lateral seismic forces applied near the top of the tower, the corresponding base overturning moments can exceed normal wind moments by as much as 255% and 336% respectively for seismic Zones III and IV. The corresponding tower drifts at the CAB level are approximately 0.30 inches due to an 80 mph wind and 1.10 inches due to lateral seismic forces in Zone IV. For concrete framed towers in seismic Zones III and IV, the inherent magnitude of lateral forces, base overturning moments and potential drift or displacement corresponds directly to an increase in wall panel thickness and/or reinforcement plus increased corner column area and reinforcement requirements as well as substantially larger foundations. Conversely, incorporation of concrete floor diaphragms normally results in a stiffer tower with less wind induced movement as well as offering self dampening characteristics. (Other advantages of a concrete framed tower include site adaptability, local material cost, aesthetic appearance, low sound attentuation and positive psychological impact on the user due to minimum movement).

The values in Table 1 above represent average design wind speeds but it should be noted that lateral wind forces can increase substantially with the square of the wind speed ($q = 0.00256\ kz\ (IV)^2)^2$. Towers sited in extremely high wind zones, such as Thule, Greenland, or Anchorage, Alaska can experience wind gusts well beyond 130 mph which equates to a base wind pressure of approximately 75 psf versus 28.6 psf for an average design 80 mph wind speed. Towers sited in regions of both high wind exposure as well as seismic Zone IV classification, such as some coastal areas of Alaska, require increased awareness and design judgment by the structural engineer.

Figure 2 represents a comparison between lateral wind and seismic forces for the same tower as Figure 1 framed with structural steel and insulated metal curtain walls. Total tower weight for comparative purposes is 1100 kips. Correspondingly, Table 2 below summaries the tower base shear and overturning forces:

Table 2

		Seismic			
	Wind	Zone I	Zone II	Zone III	Zone IV
Base Shear	136^k	25^k	51^k	101^k	134^k
Base overturning moment	$11,000^{k-'}$	$2,700^{k-'}$	$5,400^{k-'}$	$10,800^{k-'}$	$14,300^{k-'}$

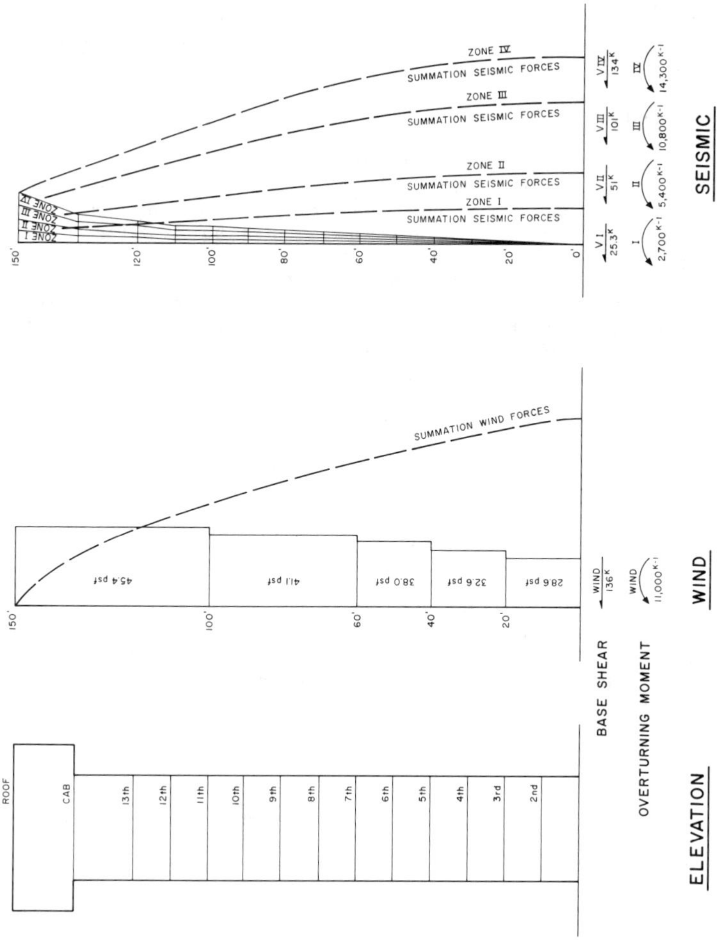

Fig. 2. LATERAL FORCES - WIND VS. SEISMIC (STEEL TOWER)

For the lighter, structural steel framed tower wind forces normally exceed seismic lateral forces with the exception of Zone IV where the total seismic overturning moments can be as much as 30% greater. Moreover, the impact of seismically induced vertical tower acceleration in combination with calculated horizontal, static forces can result in significantly higher uplift forces on the foundation in both seismic Zones III and IV. Incorporation of greater than 50% of the local tower column dead loads into the design calculations in recognition of potential vertical ground acceleration requires engineering judgment by the structural engineer.

Drift Control

Control tower drift requirements are not specifically dealt with in the current Uniform Building Code or SEAOC recommendations. For conventional building type structures both references recommend a seismic drift restriction of 0.005 x H for story to story displacement at design seismic loads and a corresponding "drift" value of 1/K x displacement. For a 150 ft. high structure this equates to a total drift of 9 inches at the top relative to the base assuming K = 1.0. The recommended wind drift limitation established by SEAOC (1980) is 0.0025 x H, or approximately 4-1/2 inches, for a height to drift ratio of H/400.

Because of the unique and critical functions that control towers serve ABKJ has set greatly restricted tower drift limits. It should be recognized that the conventional limitations are recommended design minimums allowing the structural designer to evaluate each specific building and establish a more realistic limit based on occupancy type, period, and comfort. For the majority of the control towers there is a 24 hour occupancy requirement during which the operational functions warrant absolute minimum distractions due to tower movement. A reduction in the drift limitation to H/3000 is usually inherent with concrete framed type towers. It is recommended that a similar drift limitation value shall also be attained in the case of steel-framed towers.

Tower Geometry

Selection of tower shape and occupancy influences structural stiffness. Typically the user requirements mandate either an occupied shaft, with as many as 12 to 14 floor levels, or an unoccupied shaft coupled with upper mechanical/electrical floors and CAB levels plus at-grade-structure(s) for administrative functions. Selection of occupied type shafts (as normally selected by Air Force users) with approximately 400 sq. ft. area floor levels results in plan profiles ranging from rectangular to hexagonal shapes resulting in varying degrees of shaft stiffness. Selection of unoccupied type shafts (as normally selected by FAA users) requires minimal stair and elevator framing enclosures but can result in decreased tower stiffness. Tower geometry can be influenced remarkably by seismic design requirements, such as the Edwards Air Force Base tower, where splayed hexagonal corner columns provide a path for lateral forces to be delivered directly to drilled pier type foundations.

Foundations

A relatively high tower height to base width ratio (H/D) normally results in large tower base overturning moments. In seismic Zones III & IV the magnitude of the moments increases remarkedly resulting in unusally large mat type foundations or pile or drilled pier type supports. Correspondingly, the tower base connections to the foundation are extremely important in Zones III and IV due to the magnitude of corner uplift forces. For a 150 ft. high, occupied tower with an H/D ratio of 6.5 to 1, located in seismic Zone IV, seismically induced uplift lateral forces can reach magnitudes of 1000 kips depending on the designer's assessment of potential vertical ground acceleration. Direct continuity of corner column forces to the foundation is mandatory and best achieved by direct embedment of vertical corner reinforcement and/or high strength post-tensioning bars and end plate anchorages cast monolithically with the foundation and stubbed for mechanical coupling. The usage of bolted or welded mechanical connections for concrete tower columns should be evaluated thoroughly. The usage of P/T strand splayed-end type anchorages is normally impractical due to high embedment depth requirements and should be avoided.

Selection of concrete mat type foundations versus pile-supported is a function of soil parameters, cost effectiveness and an evaluation of potential long-term foundation settlement. The magnitude of CAB level tower displacement due to a relatively small differential foundation settlement can be as much as 500% greater due to large tower height to foundation width ratios mandating a conservative approach for the foundation design. A combination of the mat foundation with grade level soil berming was utilized at Fairchild AFB (Example 2) to reduce seismic uplift forces on the piles as well as reduce excavated soil disposal quantities. Usage of belled caissons in high seismic zones can also provide increased cost effective tower overturning resistance for both uplift and compression forces.

CONSTRUCTION CONSIDERATIONS

Selection of the tower structural frame requires a thorough review of key factors affecting the constructability of the tower components. Items to consider include:

a. Site location - Local weather conditions, shipping schedule and the availability of skilled labor and materials may mandate the usage of 100% prefabricated components or entirely site-cast concrete elements.

Siting of the tower adjacent to active taxiways or runways with minimal staging may preclude the designers selection of a cast-in-place concrete frame in favor of plant-cast components or prefabricated steel elements.

b. Erection Techniques - Specific design items should include:

1. Utilization of prefabricated stair shafts for temporary bracing of the structural frame or precast wall during erection as well as convenient workmen access.

2. Utilization of staged post-tensioning for precast concrete wall panels. Termination of a proportion of the total vertical P/T forces at prescribed increments correlated to the tower overturning force levels provides for a stable erection process and results in cost effective stressing. Due to the unusual tower height to width ratio, utilization of tall precast components requires special consideration for seismic and wind bracing during the erection stage.

3. Usage of mechanical connections at panel joints provides both panel erection stability and redundancy in the panel to panel shear transfer mechanism.

4. For either site-cast or plant-cast concrete wall panels end-matching of vertical precast units is recommended to insure positive shear transfer between panels and preclude potential wind-driven water penetration through the horizontal joints. Treatment of the end-matched panels with epoxy coating during the erection process also provides a uniform bearing contact area along the entire panel joint.

5. Utilization of prefabricated, structural steel floor framing components can offer immediate diaphragm installation as well as a usable work platform for the erection crew.

DESIGN EXAMPLES

A. Anchorage International Airport Control Tower (3)

Background: The 1965 Anchorage earthquake destroyed the existing reinforced concrete frame causing the death of 2 operators. This was subsequently replaced with a low structural steel structure. In 1977 the tower was removed and replaced by a taller, precast concrete tower (Plate 1).

Tower Geometry: Square shaped, unoccupied shaft to 127 feet above grade measuring 20 ft. x 20 ft. in plan. Total tower height 169 ft. above grade plus a 13 ft. deep below grade level. (The central structural steel elevator and stair shaft is enclosed by precast, concrete wall panels x 10" thick with edge members).

Sesmic Design Criteria: Seismic Zone IV, NAVFAC P-355 and UBC (1973 edition) design codes with K = 1.0 for tower shaft and K = 2.0 for CAB framing to account for magnification at upper levels. Total seismic base shear = 290 kips at K = 1.0, Z = 1.5, T = 0.51 sec. and a base overturning moment = 67,604 Kip-ft. for K = 2.0. Design wind speed = 150 mph equates to a base shear force = 443 kips and a base overturning moment = 45,657 kip-ft. Tower overturning forces transferred through corner column fins by post-tension rods directly to foundation walls and support mat.

Plate 1. Anchorage International Airport Air Traffic Control Tower.

Foundation: Tower gravity loads and overturning forces transferred directly to below grade concrete ring beam supported by 7 ft. deep x 40 ft. sq. concrete mat with a 4'-0" diameter belled concrete pier at each corner.

Tower Drift: Theoretical CAB level drift of 0.45 inches at seismic zone IV forces (K=1.0) and 0.53 inches from 150 mph wind forces.

Construction Features: All wall panel segments precast in Seattle and transported by rail and barge to the site. Four wall panels erected in each lift each day followed by segmental, vertical post-tensioning after each three-lift stage (Plate 2). The internal structural steel stair framework was preassembled and erected prior to placement of wall panels to provide both workmen access as well as temporary wall bracing (Plate 3). All wall panels cast with end-matched joints and installed with epoxy adhesive at bearing surfaces. All post-tension tendons pressure grouted.

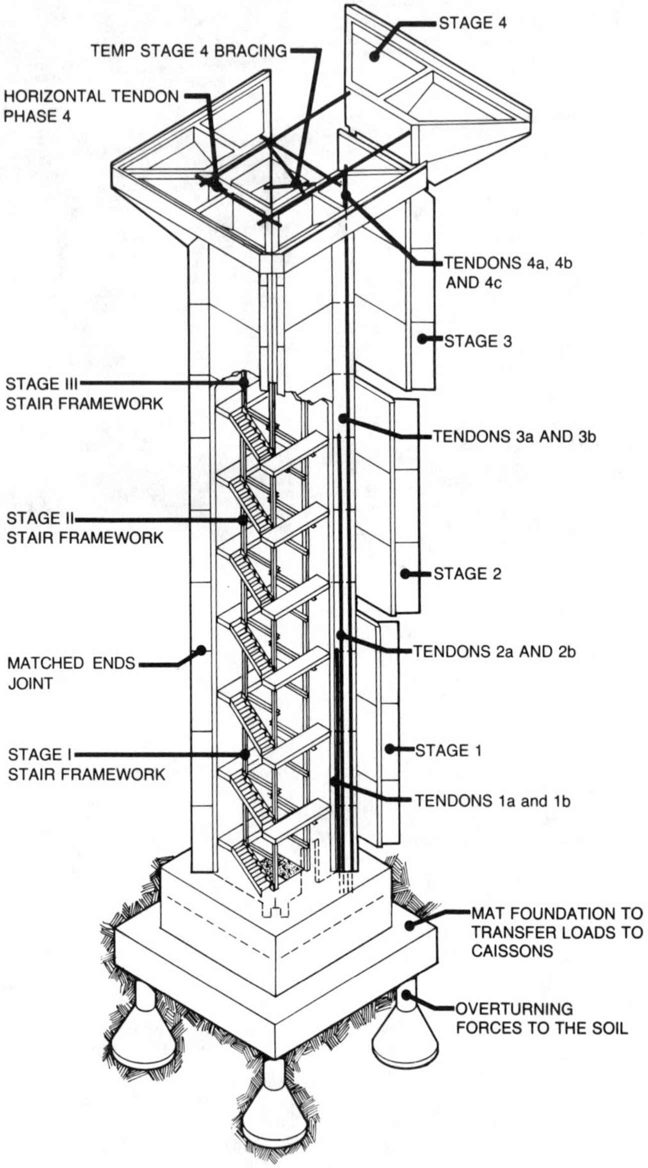

Plate 2. Anchorage International Airport Control Tower - Erection procedure and post-tensioning sequence.

Plate 3. Anchorage International Airport Control Tower - Stair framework projects ahead of precast wall panels for bracing and workman access.

Plate 4. Fairchild Air Force Base Air Traffic Control Tower near completion.

B. Fairchild Air Force Base Control Tower, Spokane, Washington

Background: The existing steel-framed control tower will be demolished subsequent to completion of a new concrete framed tower in mid-1984 (Plate 4).

Tower Geometry: Hexagon shaped, occupied shaft to 86'-8" CAB level above the first floor. Total tower height 100 ft. above entry level with nine floor levels x 24'-8" minimum width. Precast concrete wall panels segmentally cast & post-tensioned in-place with exposed concrete finish.

Seismic Design Criteria: Seismic Zone II, AFM 88-3 design code with K = 1.33, T = .51 sec., I = 1.5; CS = 0.14. Total seismic base shear = 149 kips and base overturning moment = 9821 kip-ft. Design wind speed = 100 mph equates to a base shear force = 89 kips and a base overturning moment = 4909 kip-ft. Tower overturning forces transferred directly to mat type foundation by embedded post-tension rods and anchor plates.

Foundation: Tower gravity loads and overturning forces reacted by 34'-8" hexagon shaped concrete mat 2'-6" deep supported by 18-14" HP steel piles driven to rock strata 65 ft. ± below grade. Excavated soil used for berming to increase overturning resistance.

Tower Drift: Theoretical CAB level drift of 0.14 inches from lateral seismic design forces and 0.11 inches from 100 mph wind forces.

Construction Features: All wall panel segments plant-cast in end-matched forms and trucked to site. Panel sizes to 14'-3" wide x 42'-3" long with variable thickness; 6" at midpoint to 10" at corners (Plate 5). First stage wall panels erected and braced followed by installation of pre-fabricated structural steel floor framing assembly including metal deck. Steel stair units pre-fabricated to decrease erection time followed by Stage II & III panel placement and subsequent staged post-tensioning at the panel corners. All post-tension tendons pressure grouted. Entire CAB roof framing assembly designed to be pre-assembled at grade level.

Plate 5. End-matched precast wall panels - Fairchild A.F.B. Control Tower

C. Edwards Air Force Base Control Tower, California

Background: An existing structural steel framed tower is scheduled to be replaced in 1984 by a taller steel framed tower (Plate 6).

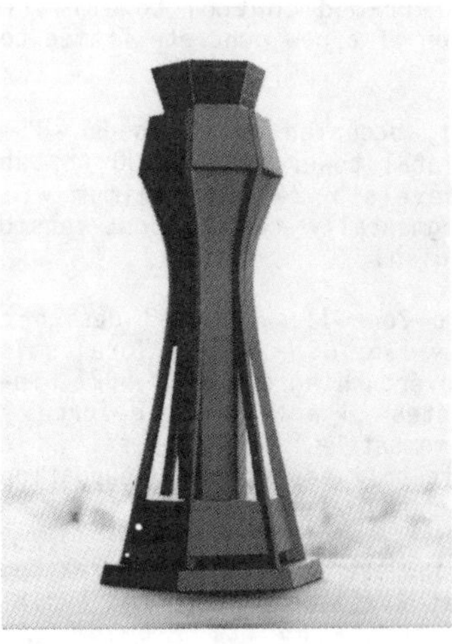

Plate 6. Model of Edwards A.F.B. Control Tower

Tower Geometry: Tapered, hexagon shaped shaft with occupied levels at grade level and the upper 4 levels. Total tower height of 150 ft. and a base width of 48 ft. across the diagonal tapering to a minimum of 25 ft. at approximately mid tower height.

Structural steel frame totally clad with Glass Fiber Reinforced Concrete curtain wall panels.

Seismic Design Criteria: Seismic Zone IV, AFM 88-3 with K = 0.67, T = 1.64 sec., I = 1.5, CS = 0.78. Total seismic base shear = 260 kips and a base overturning moment = 38,564 kip ft. at K = 2.0. Design wind speed of 120 mph per ANSI A 58.1, exposure C. Total wind base shear of 307 kips and a base overturning moment = 12,433 kip ft. Structural steel ductile moment frame designed for an equivalent K = 2.0 based on wind critical lateral forces.

Foundation: Tower gravity loads and overturning forces reacted by 5'-0" diameter drilled and belled caissons founded approximately 30 ft. below grade.

Tower Drift: Theoretical CAB level drift of 1.5 inches from lateral seismic design forces and 1.1 inches from 120 mph wind forces.

Construction Features: The central elevator and stair support shaft is designed to be shop assembled in two lengths to 84 feet maximum. All 6 corner columns prefabricated as tapered box columns (24" x 36" max.) in two lengths to 89 ft. maximum. Stage I erection includes placement of the central stair and elevator shaft followed by the lower box column sections and corresponding spandrel beams. Stage II erection allows the upper elevator core assembly to be installed followed by preassembled floor framing at each of the upper 3 levels without shoring. Floor framing assembly designed to be dropped over the core shaft, connected and cantilevered in each direction until the upper corner columns sequentially erected. Entire CAB roof framing assembly including metal decking designed to be pre-assembled at grade level and erected as a single unit.

D. Mather Air Force Base Control Tower, Sacramento, California

Background: An existing structural steel framed tower is scheduled to be replaced in 1984 by a new concrete structure. Design of the tower is a derivative from the Fairchild tower utilizing a square shape in lieu of a hexagon. Other design features include:

- Precast wall panels, segmentally post-tensioned; exposed concrete finish; end-matched casting.
- Concrete mat foundation 40 ft. square x 6'-6" deep.
- Tower overturning forces transferred directly to the mat by embedded post-tension rods and anchor plates.
- Epoxied panel joints.
- Prefabricated floor and CAB framing.

SUMMARY

Seismic design of air traffic control towers currently conforms to Federal, State and Local building codes without adequate restraints on several design criteria such as the following:

- Increased consideration of tower drift response due to lateral forces is recommended.
- Due to extremely high tower height to width ratios, most tower designs require specially designed foundations to limit potential building displacement due to differential settlements.
- Analyses of tower overturning forces should also consider a conservative assumption of potential hold-down forces as created by vertical ground accelerations.
- Due to the sensitive usage of the building the user should be assured that the tower structure be vibration free to preclude any disruption of tower communications during extreme weather conditions.

Usage of post-tensioned segmentally-cast concrete structures offers inherent tower stiffness as well as numerous advantages during the construction process. On the other hand, usage of structural steel framed towers results in lightweight construction with possible prefabrication of components. Extensive bracing is normally required to control the magnitude of lateral drift in steel towers.

References

(1) AFM 88-3, Chapter 13, equation (3-1)
(2) ANSI A58.1-1982 (Eq. 3)
(3) PCI Journal VOL. 23, No. 3; May/June 1978.

Port of Seattle
Seismic Waterfront Design

by David Owen Torseth, P. E.*

Abstract: As an owner, the Port of Seattle has its own special needs and approaches to the seismic design of its waterfront structures. The purpose of this paper is to describe the soils seismic design process, using the design developed for recently completed concrete apron project at Terminal 20 as an illustration.

The Port's standard apron section, which has been used as a basis for Port apron projects for the last twenty years, consists of a 100 foot wide concrete wharf supported by 16-1/2 inch octagonal concrete piles (see figure 1). The apron superstructure is made up of concrete precast deck panels incorporated into cast-in-place pile caps covered by crushed rock and asphaltic concrete pavement. To improve slope stability, a forty foot steel sheetpile is installed to retain the fill. Batter piles are installed along the inboard edge of the concrete apron to resist lateral forces. An elevation of -50 (MLLW) is required in the berth area. Because of the high cost of concrete platform construction, the fill slope is steepened as much as possible to decrease the amount of apron width. As a result, slope stability is normally the most critical aspect of project design. On past Port projects this normally means that the slope is built with imported select fill at 1.75 horizontal to 1 vertical. In the Terminal 20 project, though, the slope was cut to the 1.75 horizontal to 1 vertical slope and little select fill was placed. Harding Lawson Associates was Terminal 20 soils consultant and ABAM Egineers, Inc. was project structural consultant.

The soils profiles encountered at Terminal 20 are similar to other Port facilities within the local area. The native top subsurface layer consists of very soft black organic silt up to fourteen feet thick. When dredged, these materials are normally considered by environmental agencies to be unsuitable for deep water disposal and therefore require on-site pit disposal. Beneath the organic material is an alluvial deposit consisting of interlayered medium dense silty sand and medium stiff sandy silt. Silt content can vary over relatively short distances, with the less silty layers being more susceptible to liquefaction. Glacial till is normally encountered approximately 110-160 feet (35 to 50 meters) below mean lower low water. The till consists of very hard, gravelly, sandy silt with SPT N-values greater than 50 bpf.

*Project Manager, Waterfront Facility Design, Port of Seattle, P. O. Box 1209, Seattle, Washington 98111

FIGURE 1

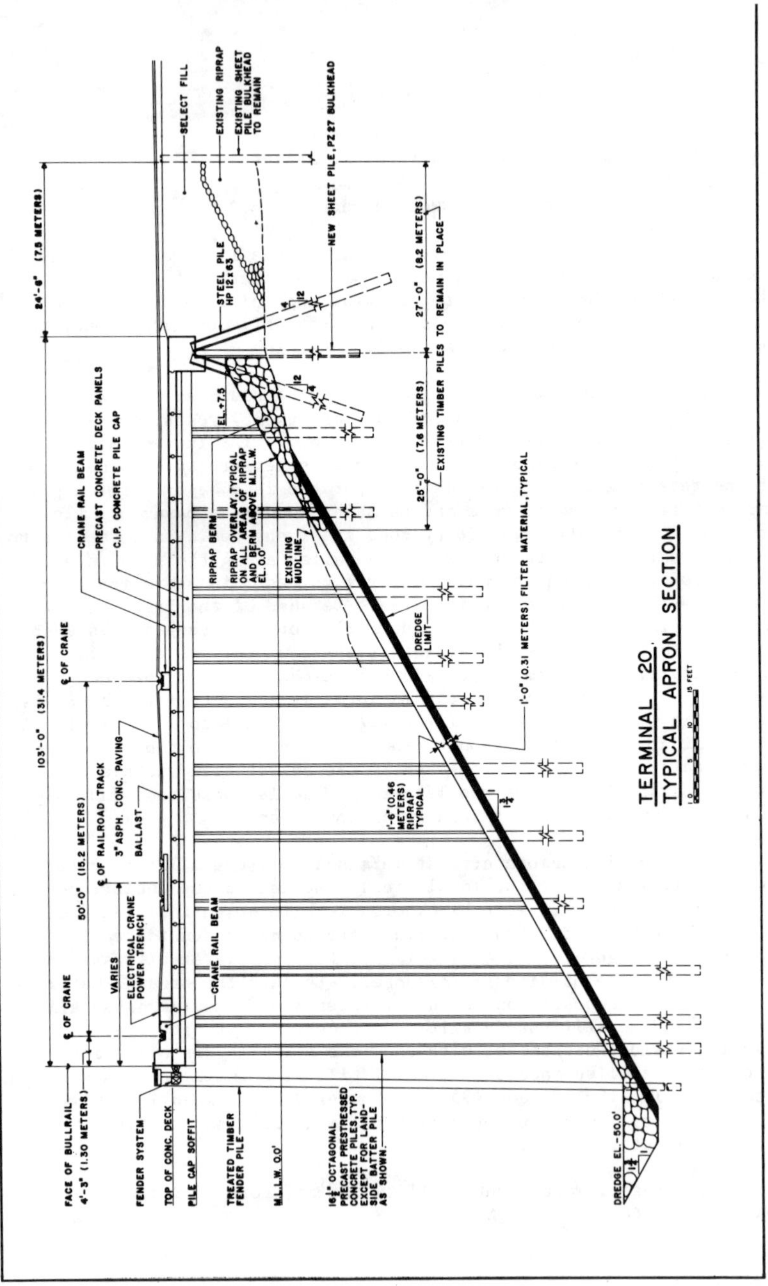

FACE OF BULLRAIL
4'-3" (1.30 METERS)
FENDER SYSTEM
TOP OF CONC. DECK
PILE CAP SOFFIT
TREATED TIMBER FENDER PILE
M.L.L.W. 0.0'
16½" OCTAGONAL PRECAST PRESTRESSED CONCRETE PILES, TYP. EXCEPT FOR LAND-SIDE BATTER PILE AS SHOWN.
DREDGE EL.-50.0'
℄ OF CRANE
VARIES
ELECTRICAL CRANE POWER TRENCH
CRANE RAIL BEAM
50'-0" (15.2 METERS)
℄ OF RAILROAD TRACK
3" ASPH. CONC. PAVING
BALLAST
103'-0" (31.4 METERS)
℄ OF CRANE
CRANE RAIL BEAM
PRECAST CONCRETE DECK PANELS
C.I.P. CONCRETE PILE CAP
RIPRAP BERM
RIPRAP OVERLAY, TYPICAL ON ALL AREAS OF RIPRAP AND BERM ABOVE M.L.L.W. EL. 0.0'
EXISTING MUDLINE
EL.+7.5
DREDGE LIMIT
1'-6" (0.46 METERS) RIPRAP TYPICAL
1'-0" (0.31 METERS) FILTER MATERIAL, TYPICAL
24'-6" (7.5 METERS)
STEEL PILE HP 12 x 63
SELECT FILL
EXISTING RIPRAP
EXISTING SHEET PILE BULKHEAD TO REMAIN
NEW SHEET PILE, PZ 27 BULKHEAD
25'-0" (7.6 METERS)
27'-0" (8.2 METERS)
EXISTING TIMBER PILES TO REMAIN IN PLACE
TERMINAL 20
TYPICAL APRON SECTION
0 5 10 15 FEET

Earthquake design considerations are considered within the context of the whole facility's analysis. The amount and type of seismic analysis required is considered on a site-by-site basis depending on the perceived site sensitivity. Since the soils response is the most crucial portion of the earthquake analysis, the Port pays special attention to subsurface investigation. For example, on the Terminal 20 project site, six deep boreholes and four additional shallow borings were taken within the seven hundred fifty by one hundred foot apron construction area. Considerable soils information is also available from neighboring Port facilities and that is also closely analyzed.

Because of the high costs associated with removal or densification of all potential liquefaction layers, the Port has not designed apron facilities to eliminate the possibility of any damage during a major earthquake. The Port has instead tried to quantify the amount of damage/risk expected during the design earthquake(s) and compare it against initial capital costs. Our past history indicates that the amount of rehabilitative work as a result of earthquake damage does not generally justify the expense of removing all liquefaction risk. The Port's risk is evaluated in light of the potential for loss of life (most important) and the potential damage to the facility and subsequent rehabilitative costs. In the case of Terminal 20, the project was designed to minimize facility damage during a earthquake which has a fifty percent probability of being exceeded during the 50 year design life. Peak ground acceleration in the glacial fill for that earthquake was estimated at 0.2g with surface peak ground acceleration being .16g. The project was additionally analyzed for an earthquake that had a ninety percent probability of not being exceeded during the 50 year design life (till peak ground acceleration on .35g with a peak ground surface acceleration of .25g). In the latter case, local reparable damage would be expected but damage would not be cataclysmic and there would be minimal danger to human life.

The Port has consistently hired soils consulting firms who have had experience in performing seismic risk analysis near the project area. Seismic risk analyses commonly seen in the Seattle area take a probabilistic approval utilizing historical earthquake data, analyzing the distance of the project site from earthquake intensive areas. Unfortunately, the Terminal 20 site, and the greater Seattle area as a whole, statistically is more seismically active than the remainder of the state.

Pseudo-static analytic methods have been consistently used by the Port for structural and slope stability design, with these methods being used on the Terminal 20 project. (Pseudo-static methods represent seismic forces as a percentage of the structure's/system's weight applied in a horizontal direction.)

Because of the ductility of the structure; because the entire structure would probably not accelerate in the same direction at the same time; and because the peak acceleration would occur for only a short time interval; the project peak acceleration was reduced to a pseudo-static force of .15g for both slope stability and structural design. The Terminal 20 structural design proceeded conventionally using that information.

Slope stability failure circle analysis was performed for the project site under three conditions: 1) during construction, 2) after construction is completed and 3) during a seismic event after construction is completed. As is shown on figure 2, the worst case occurred during the seismic event with failures beneath the existing bulkhead and the new bulkhead. Safety factors of 1.35 and 1.30 were achieved. Although these safety factors may seem low when compared to structural design safety factors (1.4 dead load and 1.7 live load) the Port does feel comfortable with them for the following reasons: 1) The failure circle analysis assumes that the peak seismic shear force will occur along the entire failure circle at the same time in the same direction. This may not be the case. 2) The effect of concrete piling driven through the slope area would be to increase slope stability. 3) Past Port slope stability experience. Some Port facilities survived the most recent major Seattle earthquake with little apparent distress even though subsequent stability analysis indicated slope stability safety factors of 1.0 or below. (These facilities have since been rehabilitated and slope stability has been improved.)

The stability analysis for the seismic case did not account for loss of soil strength due to liquefaction. Liquefaction potential was evaluated using the empirical simplified Seed method, a technique that evaluates soil cyclic shear strength using information derived from Standard Penetration Tests. These results were adjusted to allow for variations in project soils types, amount of overburden, and project slope configuration. Since project soils are so variable, these analysis were performed by looking at each individual sample for both the upper and lower level earthquakes. The results indicated that liquefaction could occur at many locations and depths within the alluvium during the more stringent design earthquake but would only minimally occur, if at all, during the lower level event.

The last major earthquake in the Seattle area (magnitude 6.5 on the Richtor scale) occurred on April 29, 1965. This earthquake is approximately equal to the lower level event assumed for design. Although many of the Port's facilities similar to Terminal 20 were built after the earthquake, there were enough concrete pile aprons with similar slopes constructed to provide the Port some indication on how their designs would behave under this seismic condition.

FIGURE 2

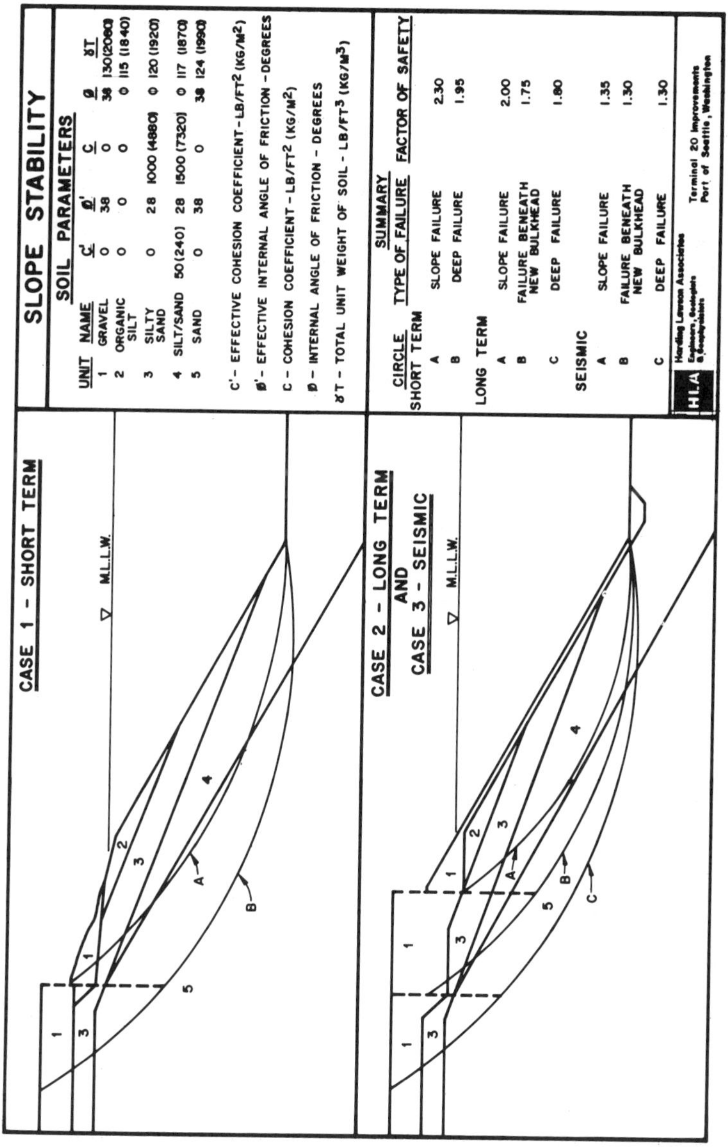
SLOPE STABILITY
SOIL PARAMETERS
UNIT NAME C' Ø' C Ø ४T
1 GRAVEL 0 38 0 38 130(2080)
2 ORGANIC SILT 0 0 0 0 115 (1840)
3 SILTY SAND 0 28 1000 (4880) 0 120 (1920)
4 SILT/SAND 50 (240) 28 1500 (7320) 0 117 (1870)
5 SAND 0 38 0 38 124 (1990)
C' - EFFECTIVE COHESION COEFFICIENT - LB/FT² (KG/M²)
Ø' - EFFECTIVE INTERNAL ANGLE OF FRICTION - DEGREES
C - COHESION COEFFICIENT - LB/FT² (KG/M²)
Ø - INTERNAL ANGLE OF FRICTION - DEGREES
४T - TOTAL UNIT WEIGHT OF SOIL - LB/FT³ (KG/M³)
SUMMARY
CIRCLE TYPE OF FAILURE FACTOR OF SAFETY
SHORT TERM
A SLOPE FAILURE 2.30
B DEEP FAILURE 1.95
LONG TERM
A SLOPE FAILURE 2.00
B FAILURE BENEATH NEW BULKHEAD 1.75
C DEEP FAILURE 1.80
SEISMIC
A SLOPE FAILURE 1.35
B FAILURE BENEATH NEW BULKHEAD 1.30
C DEEP FAILURE 1.30
HLA
Harding Lawson Associates
Engineers, Geologists & Geophysicists
Terminal 20 Improvements
Port of Seattle, Washington
CASE 1 - SHORT TERM
M.L.L.W.
CASE 2 - LONG TERM AND CASE 3 - SEISMIC
M.L.L.W.

The only significant damage occurred at Terminal 5, where a 60 foot wide concrete apron is supported by concrete piles. Lateral forces at that site are supported by brace piles at the steel sheetpile. A fill slope beneath the apron is 1.5 horizontal to 1 vertical.

Damage included subsidence of the fill behind the steel sheetpile bulkhead, upward movement of the sheetpile cap above the sheetpile bulkhead, seaward displacement of up to ten inches of concrete apron, and the cracking of concrete pile caps within ten feet of the bulkhead. Although weaker alluvial material was present deep beneath the facility, a deep-seated failure was not considered to have caused the problem. Analysis concluded that seismic loadings had overloaded the bulkhead and had caused the outboard movement of the apron and bulkhead. In addition, liquefaction in the upper portion of the slope probably resulted in a loss of passive pressure on the bulkhead. The brace piles prevented the sheetpile and cap from moving laterally like the rest of the facility and caused it to therefore move upward. Slumping of the backfill behind the bulkhead was a result of the bulkhead movement.

It is difficult to develop economical, effective measures that will mitigate earthquake damage because of the effects of the deep alluvial material. For the Terminal 20 project, though, the following construction activities were considered.

(a) Dredging out poorer underlying soils and replacing with better material.

(b) Leaving in-place existing timber piles from the pre-existing timber apron.

(c) Using a less steep slope as a basis for design of the fill slope with the concrete apron.

Dredging the poorer underlying materials from beneath the apron and replacing with better material was considered as a method for providing a better foundation for the slope, but disposal considerations limited the amount of material that could be removed. Within the Seattle area, the environmental dredged material suitability dictates the methods allowed for disposal. Chemical analysis determined that much of the Terminal 20 project dredge materials were PCB laden and would therefore require expensive upland pit disposal with a minimum of five feet of select fill coverage on top. Only one site within the Seattle area was found that could accommodate even a limited amount of the material. As a result, dredging was minimized so that material could be placed within that pit.

Leaving existing piles in-place and driving new displacement piles was considered a simple method of improving the densification of the potentially liquifiable layers. These piles reduce the possibility of local liquefaction and increase slope stability, although they make it difficult for the contractor to maneuver his equipment while constructing the facility and therefore increase project costs. Existing timber piles were left in-place within 25 feet of the new sheetpile bulkhead to provide additional stability for the bulkhead, as a compromise between project constructibility/cost and increased safety.

Use of a less-steep slope would provide additional slope stability during construction at a sacrifice of increased project costs for concrete platform and fill. Studies have indicated that the initial shear stresses in the steeper design slopes decrease liquefaction potential. As a result, the design slope is a compromise balancing project costs, slope stability, and liquefaction potential and its consequences. Based on the analysis of the local site conditions and materials, it was determined that the 1.75 horizontal to 1 vertical slope was appropriate for this project.

References

Harding Lawson Associates, Geotechnical Investigation, Terminal 20 Improvements, 300-120th Avenue NE, Bldg. 4, Suite 219, P.O. Box 3885, Bellevue, Washington 98009. March 17, 1982. (Special thanks to Glen Mannand, John Newby.)

Shannon and Wilson, Inc., Geotechnical Investigation, Terminal 30 Apron, 1105 North 38th Street, Seattle, Washington 98103. August 9, 1983.

Neil H. Twelker Associates, Structural Damage at Pier 5, West Waterway, During Earthquake of April 29, 1965, 5645-42nd West, Seattle, Washington 98119, February 7, 1966.

SEISMIC RESPONSE OF PORT AND HARBOR FACILITIES

by S.D. Werner,* M. ASCE and S.J. Hung,* M. ASCE

ABSTRACT

This paper summarizes results from a recent research program that evaluated the seismic response characteristics of port and harbor facilities. In particular, the program compiled extensive information regarding the behavior of port and harbor facilities during past earthquakes, assessed current seismic design provisions for such facilities, and evaluated the applicability of dynamic analysis as a means for enhancing these design provisions. Principal findings of the program were: (a) the potential for significant damage to port and harbor facilities is most strongly related to porewater pressure buildup in the surrounding soil materials; (b) current seismic design provisions for such facilities often fail to take proper account of many of the possible effects of earthquake ground motion; and (c) dynamic analysis can enhance the earthquake-resistant design of port and harbor facilities and should be incorporated to a much greater extent into their seismic design provisions.

INTRODUCTION

An important component of regional, national, and world-wide lifeline systems is port and harbor facilities. Such facilities function to load, unload, ship, and store cargoes of materiel and personnel, and to provide repair and maintenance support of these operations. The interruption of this vital function, for any reason, could result in serious hardships and far-reaching economic consequences.

Many port and harbor facilities are located in highly seismic regions, and it is particularly important that such facilities be engineered so that a major earthquake will not cause extensive damage and disruption of operations. In view of this, a multiyear research program has been carried out to investigate how port and harbor facilities respond to strong earthquake motions, and how they can be designed and analyzed to resist these seismic effects. In particular, this program has addressed the following issues:

- How have port and harbor facilities been damaged during past earthquakes, and what lessons can be learned from these damage observations?

- How are port and harbor facilities currently designed to resist earthquakes, and what are the merits and limitations of these design procedures?

*Agbabian Associates, El Segundo, California, USA

- How can dynamic analysis be used to enhance our ability to design port and harbor facilities against earthquake ground shaking effects?

This paper summarizes the findings of our research program pertaining to each of these issues. It is extracted from our program final report in which each issue is addressed in considerable detail [51]. It is noted that these findings pertain only to the effects of ground shaking on the behavior of the structures and soil materials that typically comprise ports and harbors. The additional effects of earthquake-induced tsunamis, which can also be important for port and harbor facilities, were not addressed in our research program.

LESSONS FROM PAST EARTHQUAKES

The first task in our research program consisted of a compilation and assessment of all available documented information regarding how port and harbor facilities have behaved during past earthquakes, and the principal causes and modes of any damage that may have occurred. This information was primarily from Japan, although significant information was also obtained from the United States and Chile.

Table 1 summarizes results from this task. It clearly shows that port and harbor facilities are susceptible to extensive damage from earthquakes. For example, prior earthquakes are seen to have caused substantial lateral sliding, deformation, and tilting of quay walls and sheet-pile bulkheads. This damage has often been accompanied by anchor-system failure and extensive settlement and cracking of paved aprons. In addition, docks and piers have undergone extensive sliding and buckling, bending, and yielding of pile supports. Evaluation of this damage indicates the following trends [9, 28, 31, 34, 48]:

- By far the most significant source of earthquake-induced damage to port and harbor facilities has been porewater pressure buildup in the loose-to-medium-dense, saturated, cohesionless soils that prevail at port and harbor sites. This has led to damage due to excessive lateral pressures applied to quay walls and bulkheads by backfill materials (Item A in Table 1), and to liquefaction (Item B), localized sliding (Item C), and massive submarine sliding of the site soil materials (Item D).

- To underscore the importance of these porewater pressure effects, several examples can be cited involving similar port and harbor structures located on adjacent sites, in which one site experienced significant porewater pressure buildup during a given earthquake and the other did not. Invariably, the structures at the sites with significant porewater pressure buildup suffered severe damage, whereas the damage to structures at the other sites was much less substantial. In other examples (from the Niigata, 1964, and Alaska, 1964 earthquakes), widespread liquefaction and massive soil sliding caused by porewater pressure buildup completely destroyed entire port and harbor areas.

TABLE 1. SUMMARY OF EARTHQUAKE-INDUCED DAMAGE TO PORT AND HARBOR FACILITIES

Earthquake			Port Location	Damage	
Location	Date	Magnitude		Description	Possible Cause(s)
Kanto, Japan	Sep 1, 1923	8.2	Yokohama and Yokosuka	Concrete block quay walls: sliding, tilting, and/or collapse with some bearing capacity failure of rubble-stone foundation Steel bridge pier: buckling of pile supports	A C,E
Kitaizu, Japan	Nov 26, 1930	7.0	Shimizu	Caisson quay wall (183 m long): tilting, outward sliding (8.3 m), and settlement (1.6 m) L-Shaped block quay wall (750 m long): outward sliding (4.5 m) and settlement (1.2 m)	A,B,C
Shizuoka, Japan	Jul 11, 1935	6.3	Shimizu	Caisson quay wall: outward sliding (5.5 m) and settlement (0.9 m) accompanied by anchor system failure	A,B,C
Tonankai, Japan	Dec 7, 1944	8.3	Yokkaichi Nagoya Osaka	Pile-supported concrete girder and deck: outward sliding (3.7 m) accompanied by extensive soil sliding Sheet-pile bulkhead with platform: outward bulging (4 m) Steel sheet-pile bulkhead: outward bulging (3 m)	A,B,C
Nankai, Japan	Dec 21, 1946	8.1	Nagoya Yokkaichi Osaka Uno	Sheet-pile bulkhead with platform: outward bulging (4 m) Pile-supported concrete girder and deck: outward sliding (3.7 m) Steel sheet-pile bulkhead: outward bulging (3 m) and settlement (0.6 m) Gravity-type concrete block and caisson quay wall: seaward sliding (0.4 m) accompanied by soil sliding	A,B,C
Tokachi-Oki, Japan	Mar 4, 1952	8.1	Kushiro	Concrete caisson quay wall: tilting, outward sliding (6 m), and settlement (1 m)	A,B,C
Chile	May 22, 1960	8.4	Puerto Montt Talcuhuano	Concrete caisson quay walls: overturning and extensive tilting Steel sheet-pile seawall: outward sliding (up to 1 m) and anchor failure Gravity-type concrete seawall: complete overturning and sliding (1.5 m) Concrete block quay wall: outward tilting	A,B,C A,B
Alaska	Mar 27, 1964	8.4	Anchorage Valdez Whittier Seward	Dock structures: extensive seaward tilting with bowing, buckling, and yielding of pile supports Entire harbor: destroyed by massive submarine landslide Pile-supported piers and docks: buckling, bending, and twisting of steel pile supports Steel sheet-pile bulkhead: extensive bulging Major portion of harbor: destroyed by massive submarine landslide	B,D,E B,D

Note : Table continued on following page.

TABLE 1. (CONTINUED)

Earthquake				Damage	
Location	Date	Magnitude	Port Location	Description	Possible Cause(s)
Niigata, Japan	Jun 16, 1964	7.5	Niigata	Extensive damage due to liquefaction and sliding of soil strata. Summary of damage is as follows: Piers and landings: sliding (up to 5 m), submergence, and tilting Sheet-pile bulkheads: sliding (over 2 m), submergence, settlement (up to 1 m), and tilting. Extensive anchor failure Quay-walls: sea sliding (up to 3 m) settlement (up to 4 m) with extensive anchor failure and wall tilting	A,B,C
Tokachi-Oki, Japan	May 16, 1968	7.8	Hachinohe	Steel sheet-pile bulkheads: outward sliding (0.9 m), tilting, and settlement, with anchor failure	A
			Aomori	Gravity-type quay wall: sliding and settlement (0.4 m) Gravity-type breakwater: sliding (0.9 m) and pavement settlement (0.9 m)	A
			Hakodate	Steel sheet-pile bulkhead: seaward tilting (0.6 m) and apron settlement (0.3 m) Quay-wall: settlement (0.6 m) and sliding (0.4 m)	A,B
Nemuro-Hanto-Oki, Japan	Jun 17, 1973	7.4	Hanasaki	Gravity-type quay wall: sliding (1.2 m) and settlement (0.3 m) with corresponding apron settlement (1.2 m) Steel sheet-pile bulkhead: sliding (2 m) and anchor failure	A,B
			Kiritappu	Steel sheet-pile bulkhead: relatively minor damage Gravity-type quay walls: relatively minor damage	
Miyagi-Ken-Oki, Japan	Jun 12, 1978	7.4	Shiogama	Concrete gravity-type quay wall: outward tilting (0.6 m) and apron pavement settlement (0.4 m)	A,B
			Ishinomaki	Steel sheet-pile bulkheads: outward sliding (up to 1.2 m) and apron settlement (up to 1 m) Concrete block retaining wall: sliding, tilting, and cracking with corresponding pavement settlement (0.2 m) relative to wall	
			Yuriage	Concrete block gravity quay wall and steel sheet-pile bulkhead: large horizontal displacements (up to 1.2 m)	
			Sendai	Steel sheet-pile bulkheads: cracking and settlement of apron and pavements	

Legend

A: Excessive lateral pressure from backfill materials, in the absence of complete liquefaction, and possibly accompanied by reduction in water pressure on outside of wall
B: Liquefaction
C: Localized sliding
D: Massive submarine sliding
E: Vibrations of structure

- There is very little evidence of damage directly related to the earthquake-induced vibrations of the structures themselves (Item E in Table 1). This may be either because direct structural effects are overshadowed by the effects of porewater pressure buildup in the adjacent soils, or because the seismic design provisions related to structure vibratory effects are adequate or conservative. The use of dynamic analysis to evaluate the earthquake-induced behavior of port and harbor facilities could provide important insights along these lines.

- Insights into the importance of the seismic coefficients used in the design of port and harbor facilities are provided by observations from Japanese earthquakes. For example, in the earthquakes at Niigata (1964), Tokachi-Oki (1968), and Nemuro-Hanto-Oki (1973), affected port and harbor facilities were all designed using seismic coefficients whose values were about half the values of the peak accelerations actually experienced during the earthquakes. Despite this similarity, these three earthquakes caused widely varying degrees of harbor damage. This damage appeared to be most closely related to the extent of liquefaction of the affected harbor sites, which was greatest during the Niigata earthquake and was least during the Tokachi-Oki event. This suggests that the seismic coefficients used in design of port and harbor facilities are of secondary importance, when compared to the potential for liquefaction of the site soil materials.

- Although the effects of porewater pressure buildup and liquefaction are predominant, the structure configuration including the anchor system used for retaining wall structures also plays an important role in the behavior of port and harbor facilities during earthquakes. For example, the failure or survival of sheet-pile bulkhead structures has been strongly influenced by both the placement and the type of anchor system used (see further discussion of sheet-pile bulkheads later in this paper). In addition, there is evidence that damage to several waterfront structures during the 1964 Niigata earthquake was probably related to pre-earthquake modifications of these structures to circumvent prior large-scale ground subsidence. These modifications involved an increase in the height of the structures, without a corresponding strengthening of the foundation to accommodate the effects of the accompanying increased structure mass and the raised center of mass.

SEISMIC DESIGN PROCEDURES

Upon establishment of the predominant factors that led to port and harbor damage during past earthquakes, the second task of this research program assessed how port and harbor facilities are designed against these factors, and whether these design procedures are adequate. This assessment was based on thorough reviews of available design-related documentation from Japan [16, 18, 19, 21, 30, 32, 34] and the United

States [4, 7, 36, 37, 41-43, 45, 49, 52], and on discussions with engineers from port authorities, government, universities, and consulting firms who are involved in the design of ports and harbors to resist earthquakes. Results of this assessment are categorized as corresponding to geotechnical considerations and structure-specific considerations.

Geotechnical Considerations

The geotechnical phenomena addressed in this seismic design assessment pertain to (a) lateral earth pressures for retaining wall structures; (b) hydrodynamic seawater pressure on retaining wall structures; (c) earthquake effects on bearing capacity; (d) seismic resistance of piles; (e) earthquake-induced slope instability; and (f) liquefaction. The design practice for ports and harbors that pertains to these phenomena is summarized in Table 2. This table indicates that most of these phenomena are either ignored or are treated using simplified pseudostatic techniques that have important limitations when used to represent dynamic effects from earthquakes. Dynamic analysis is now routinely carried out only for liquefaction investigations under free-field conditions; however, even this is a relatively recent development and a

TABLE 2. GEOTECHNICAL-RELATED SEISMIC DESIGN PRACTICE FOR PORT AND HARBOR FACILITIES

Item	Predominant Design Practice
Lateral Earth Pressures for Retaining Wall Structures	Mononobe-Okabe method most typically used, occasionally incorporating suggested simplifications by Seed and Whitman [43].
Hydrodynamic Seawater Pressure on Retaining Wall Structures	Usually ignored. Westergaard [52] procedure occasionally used.
Earthquake Effects on Bearing Capacity	Usually ignored.
Seismic Resistance of Piles	Usually either ignored or considered using pseudostatic procedures.
Earthquake-Induced Slope Instability	Pseudostatic methods most typically used.
Liquefaction	Standard procedures described by Seed [41] most typically used, incorporating empirical field performance correlations and/or one-dimensional total stress dynamic analysis techniques. Occasional use of one-dimensional effective stress methods.

potential major problem exists at many ports and harbors because possible liquefaction had been essentially ignored in past seismic design practice. In fact, when viewed in the context of the overall design requirements for port and harbor facilities, the geotechnical-related seismic effects summarized in Table 2 do not play nearly as major a role in the design process as would be desirable, in view of the extensive damage to such facilities that has occurred during prior earthquakes. Further discussion of seismic design procedures pertaining to the above phenomena is provided in the paragraphs that follow.

Lateral Earth Pressures for Retaining Wall Structures. The Mononobe-Okabe equation is the predominant method used to define earthquake-induced lateral pressures for quay walls, sheet-pile bulkheads, and other port and harbor retaining wall structures [27, 33]. This approach assumes that (a) the wall moves sufficiently to mobilize the minimum active pressure; (b) when the minimum active pressure acts against the wall, a wedge-shaped soil mass is at the point of incipient failure with the maximum shearing resistance mobilized all along the plane sliding surface; and (c) the soil wedge acts as a rigid body with earthquake accelerations acting uniformly throughout the wedge. Based on these assumptions, Mononobe and Okabe used the Coulomb sliding wedge method to obtain an expression for the total pseudostatic horizontal earthquake force as a function of the horizontal and vertical ground accelerations, and various parameters related to the soil, wall, and backfill. (Seed and Whitman [43] have since used sensitivity studies of these parameters to develop a simplified form of the Mononobe-Okabe equation that is now often used in current engineering practice.) For application to port and harbor facilities, the Mononobe-Okabe approach is limited in that it cannot represent effects of porewater pressure buildup, the possible presence of cohesive soil materials, the flexibility of the retaining wall (which can be important for sheet-pile bulkheads and even for certain quay wall configurations), and dynamic phenomena due to vibrations of the soil and structure. Although more recent procedures have been developed that circumvent some of these limitations [e.g., 40, 46], the Mononobe-Okabe approach is still the most universally used for port and harbor applications.

Hydrodynamic Seawater Pressure on Retaining Wall Structures. There is evidence from past earthquakes that hydrodynamic pressure exerted by seawater on retaining wall structures affects their response to earthquake motions; for example, failures of numerous quay wall structures have been attributed to a reduction in these water pressures [34, 43]. Despite this, dynamic seawater pressure effects are usually ignored in current design practice for port and harbor facilities. When such pressures are considered, they are typically represented as a single effective mass through the approach of Westergaard [52] which was developed for the case of rigid and infinitely long dam structures, harmonic horizontal ground motions oriented normal to the dam axis, and negligible surface waves. Evaluations of this approach have shown that: (a) this use of a single effective mass (shown by Westergaard to be frequency-dependent) does not fully describe hydrodynamic response to earthquakes, which are typically wide-band processes; and (b) Westergaard's approach is valid only for excitation frequencies less than the fundamental frequency of the water body [5]. In recent years, important progress toward providing a more complete representation of these dynamic water pressures has been made using substructured dynamic analysis procedures [e.g., 6, 15].

Earthquake Effects on Bearing Capacity. During an earthquake, the bearing capacity of the soil underlying gravity-type structures at ports and harbors may be affected in three ways. First, because of the horizontal components of the seismic excitations, the structure may undergo overturning rotation that could increase the soil pressures under its toe, to an extent where they could exceed the ultimate bearing capacity of the soil. Second, porewater pressure buildup in loose, saturated cohesionless soils could reduce their bearing capacity even if no rotation of the structure takes place. Finally, for unsaturated sands (or saturated clays) not subject to porewater pressure buildup, strain-rate effects could result in a slight increase in bearing capacity. In general, such seismic effects on bearing capacity are seldom considered directly, although effects of porewater pressure buildup are considered indirectly through liquefaction assessments. When such effects are considered, they are generally represented from engineering judgment as a slight increase or decrease relative to static conditions, depending on the particular soil conditions at the site.

Seismic Resistance of Piles. In the event of an earthquake, pile supports of port and harbor facilities function to mobilize the underlying deep (and typically more competent) soil deposits to resist the seismic forces induced by structural vibrations of piers or by lateral soil pressures applied to bulkheads and quay walls. Porewater pressure buildup in the near-surface soil deposits or in backfill may substantially increase these seismic forces, and may reduce the effective embedment depth of the pile supports. Relatively recent procedures are available to accommodate porewater pressure and dynamic effects in the seismic design of piles [e.g., 11]; however these effects are seldom represented in pile design practice for port and harbor facilities, which typically either ignores seismic phenomena or represents them as simple pseudostatic loads. Both batter piles and vertical piles are used at port and harbor facilities. Batter piles are desirable as support at anchors for sheet-pile bulkheads, because of their ability to resist lateral displacements of the sheet pile due to porewater pressure buildup during earthquakes. However, experience from past earthquakes has shown batter piles to be less desirable at piers where, because of their significant lateral stiffness, they transmit large lateral seismic forces that have resulted in severe damage to pile caps and decking [24, 28]. Vertical piles, which are more flexible laterally, are therefore more suitable as pier foundation supports.

Earthquake-Induced Slope Instability. The evaluation of slope instability can be important to the design of such port and harbor facilities as gravity-type quay walls or trestle-type piers with small retaining walls. Current seismic design practice for such facilities most typically uses a pseudostatic approach. In this approach, the stability of a potential sliding mass is determined as for static loading conditions, and earthquake effects are represented as an additional pseudostatic horizontal force acting on the sliding mass. This additional force is expressed as the product of the weight of the sliding mass under consideration and a seismic coefficient that is normally based on the seismicity of the region. Typically, these seismic coefficients range from 0.1 to 0.15. This pseudostatic approach can be useful in evaluating the performance of embankments constructed of soils that do not lose significant strength during earthquakes (e.g., clayey soils, dry sands, extremely dense saturated sands); however, there is substantial evidence that the approach cannot reliably predict potential earthquake-induced slope failures for the loose, saturated sands that typically exist at port and harbor sites and tend to lose strength due to porewater pressure buildup during strong ground shaking [42]. Alternative procedures involving dynamic analysis are available that are more directly applicable to port and harbor sites under dynamic seismic conditions but are seldom used in current seismic design practice [29, 42]. The most complete of these approaches, as described by Seed [42], involves computation of initial static and dynamic embankment stresses using finite element analyses, and application of these stresses to embankment soil samples in cyclic laboratory tests in order to evaluate the potential for slope instability.

Liquefaction. Of the various geotechnical considerations for seismic design of port and harbor facilities, the evaluation of earthquake-induced liquefaction for level ground conditions is probably the most important. For this reason, liquefaction assessments using dynamic analysis techniques are now routinely implemented at port and harbor sites although, as noted earlier, this is a relatively recent development. These assessments are typically carried out for free-field conditions, using one-dimensional total stress or effective stress models that compute the response of horizontally layered sites to vertically propagating shear waves. For certain port and harbor facilities, effects of soil/structure interaction and irregular topography may be important; for this reason, the following section of this paper (Dynamic Analysis Procedures) addresses the use of two- or three-dimensional models that incorporate these effects in dynamic analyses. In current design practice, critical stress levels leading to liquefaction are evaluated through laboratory cyclic load tests on representative undisturbed soil samples at various confining pressures [10, 41], and/or through empirical correlations of field performance observations during past earthquakes that involve some measurable in situ soil properties - usually Standard Penetration Test (SPT) blowcounts [45]. Both approaches have important benefits: the laboratory tests consider the behavior of the actual soil materials at the site, while the field performance correlations can provide valuable guidance from records of past experience. For this reason, it would seem prudent to apply both approaches at port and harbor sites. However, when applying the approaches, care must be taken to interpret their inherent uncertainties, which arise from sample disturbance (in the case of laboratory tests) or from the use of SPT blowcounts to characterize the liquefaction resistance of the soil (in the field performance correlations).

Structure-Specific Considerations

This section summarizes seismic design considerations for three important types of port and harbor structures - gravity-type quay walls, sheet-pile bulkheads, and piers.

Gravity-Type Quay Walls. Two seismic forces are considered in the design of gravity-type quay walls: the inertia force of the wall and the resultant force from pseudostatic lateral soil pressures computed from the Mononobe-Okabe equation. When subjected to these forces, quay walls are designed to avoid overturning and excessive sliding and tilting, although some horizontal wall movement is tolerated. Earthquake-induced slope instability is an important design consideration. Effects of lateral soil pressure are most important for caisson-type gravity quay walls because of their height. Remedial measures used to enhance the seismic resistance of quay walls include grouting of the backfill, widening the base of the wall, using pile supports, partially embedding the wall, and replacing underlying weak soils with more competent soil materials.

Sheet-Pile Bulkheads. Seismic design of sheet-pile bulkheads is generally based on seismic forces corresponding to pseudostatic lateral soil pressures computed from the Mononobe-Okabe equation. Designs to resist such lateral seismic forces are directed toward insuring an adequate factor of safety against toe failure, adequate bending resistance of the sheet pile, adequate tensile resistance of the tie rod, and adequate anchor resistance. Anchor design considerations include (a) placement of anchors at sufficient depths below the ground surface to avoid weak near-surface soil deposits; (b) enhancement of anchor capacity by using deeply embedded piles or sheet piles; and (c) insuring an adequate distance between the anchor and the bulkhead wall, to avoid loss of passive pressure resistance of the anchor.

Piers. Piers are generally designed to resist seismic lateral forces in the form of pseudostatic forces analogous to those used for building design. The relatively lightweight, pile-supported pier structures are less prone to seismic damage than are quay walls or bulkheads. The use of batter piles has resulted in damage to pier pile caps and decking, because of the large lateral stiffness of such piles; therefore, vertical pile supports are generally preferred.

DYNAMIC ANALYSIS PROCEDURES

Overview

The seismic design of many major structures (e.g., dams, bridges, power plants, high-rise buildings) has routinely incorporated modern dynamic analysis techniques as an integral part of the design process. For such structures, dynamic analysis has been the basis for assessing the integrity of the structure under the prescribed ground shaking, as well as for evaluating the response and stability of the surrounding soil medium. It would seem that dynamic analysis procedures should fulfill a similar function for port and harbor facilities, in view of their susceptibility to extensive earthquake damage, the economic consequences of such damage, and the limitations in their current seismic design provisions.

The use of dynamic analysis in the seismic design of port and harbor facilities involves six main steps: (a) development of suitable seismic input motions; (b) measurement of reliable soils data; (c) selection of an appropriate analysis procedure; (d) development of a suitable model of the soil/structure system; (e) implementation and interpretation of the dynamic response calculations; and (f) utilization of the analysis results in the design process. Sound engineering judgment must be exercised during each stage of such an analysis. It is particularly important to check that the results obtained and conclusions drawn are consistent with engineering experience and practice.

Keeping this overall analysis process in mind, the third task of this research program addressed one aspect of the process - the analytical procedures appropriate for application to port and harbor facilities. In particular, three general types of methods - deterministic total stress methods, deterministic effective stress methods, and probabilistic methods - were evaluated with regard to their ability to satisfy the following two requirements for port and harbor facility applications.

1. Structural Response. The earthquake-induced stresses and deformations in the structure must be determined so as to guide the development of a suitable structure design. This requires use of an appropriate soil/structure interaction analysis that accounts for the site topography and material properties and the geometry and properties of the structure.

2. Site Response. The potential for significant porewater pressure buildup and its related effects (e.g., widespread liquefaction and slope failure) must be determined so as to guide assessment of the integrity of the site under earthquake ground shaking. This requires a seismic analysis of the site soil behavior that accounts for the presence of the structure and the material properties.

These evaluations are summarized in the remainder of this section.

Deterministic Total Stress Methods

Deterministic total stress methods involve dynamic analyses that compute the total state of stress in the soil. As applied to ports and harbors in accordance with Requirements 1 and 2, such methods would be comprised of the following steps: (a) an analysis of the dynamic response of the soil/structure system; (b) conversion of the resulting irregular shear stress histories in the soil to equivalent numbers of uniform stress cycles; (c) determination of the cyclic shear stress levels at various depths necessary to cause liquefaction in the same number of stress cycles as was determined from Step b; and (d) comparison of the cyclic shear stresses induced by the ground shaking (Step b) to the critical shear stresses that cause liquefaction (Step c), in order to to identify any critical regions in the soil where liquefaction may occur. The dynamic analysis technique applied in Step a above would most typically use two-dimensional or three-dimensional finite element codes with equivalent linear or nonlinear soil models, in order to compute the structural and soil response including effects of soil/ structure interaction, the site topography, and the layering of the subsurface soil materials [1, 22, 23]. For site geometries that can be represented as horizontal soil layers subjected to vertically propagating shear waves, such analyses could be supplemented by separate computations of free-field response using one-dimensional techniques [39]. The conversion of the resulting irregular shear stress histories in the soil to equivalent shear stress cycles, as applied under Step b, can be carried out using empirical procedures based on laboratory test data [44]. Subsequent assessment of the critical cyclic shear stresses leading to liquefaction, as required under Step c and discussed earlier in this paper, involves laboratory cyclic load tests on undisturbed samples at different confining pressures [10, 41] and/or use of empirical correlations of field performance data from prior earthquakes [45].

Deterministic total stress methods, as summarized above, have been successfully and widely used to analyze the response of earth dams and embankments, and are applicable to port and harbor facilities as well. In fact, they are the only methods now sufficiently developed to satisfy Requirements 1 and 2 above for seismic analysis of port and harbor facilities. However, a limitation of total stress methods is that the dynamic analysis of the soil/structure system response (under Step a) does not incorporate porewater pressure effects or effects of uncertainties in the seismic input or soil properties. For this reason, considerable research has been directed toward the development of deterministic effective stress methods and probabilistic methods, although neither of these methods is yet sufficiently developed to be applicable to port and harbor facilities in accordance with Requirements 1 and 2. The methods do, however, represent promising dynamic analysis procedures for the future, and are summarized in this context in the remainder of this section.

Deterministic Effective Stress Methods

Deterministic effective stress methods are based on the premise that the deformations of the soil medium are controlled by its effective state of stress [2, 47]. As applied to port and harbor facilities in accordance with Requirements 1 and 2, such methods would involve the following steps: (a) laboratory tests of undisturbed soil samples to define the material parameters for the effective stress soil model; and (b) a dynamic analysis that uses this soil model in the computation of the soil/structure system response including coupled effects of porewater pressure buildup and dissipation. This incorporation of porewater pressure effects in the dynamic analysis is particularly important for port and harbor facilities, because the geometry of their adjacent backfill and subsurface soil deposits is typically such that drainage paths are relatively shorter than, for instance, those of large earth dams. It is therefore more important that the effects of dissipation of porewater pressure, as well as the buildup of porewater pressure, be considered in the analysis of such facilities. In adddition, as the strength and duration of the ground shaking increase, the influence of porewater pressure effects on the response of the saturated cohesionless soil deposits that often comprise port and harbor facilities will also increase. Therefore, properly conceived and applied effective stress methods should provide improved estimates of the port and harbor facility response, when such methods are further developed as discussed below so as to satisfy Requirements 1 and 2.

In current effective stress methods, the strain dependence of the soil material properties is now considered using nonlinear material models or, less commonly, equivalent linear models. In addition, various methods for coupling the behavior of the soil matrix and the porewater are employed, ranging from loosely coupled techniques (which use separate models to describe dynamic mechanical behavior and porewater flow [e.g., 3, 12, 25]) to fully coupled techniques (which treat the soil and porewater as a two-phase mixture in which all interactions between mechanical and porewater flow phenomena are modeled [13, 20, 55]). However, these various methods are now primarily available only as one-dimensional procedures for analyzing liquefaction potential under free-field conditions at sites comprised of horizontal soil layers and subjected to vertically propagating shear waves. Very few two-dimensional or three-dimensional effective stress methods necessary to perform soil/structure interaction analyses of port and harbor faciities (in accordance with Requirement 1) have been developed, and these are not yet readily available. Furthermore, the empirical soil/fluid parameters that comprise many current effective stress models are often not readily defined from conventional cyclic testing procedures.

The desirability of extending existing effective stress methods to accommodate two-dimensional or three-dimensional conditions and to provide an improved experimental basis for identifying the soil/fluid model parameters is recognized by numerous geotechnical engineers, and work is proceeding in these areas. Such extensions should contribute substantially to our understanding of the seismic response of port and harbor facilities, as well as other structures (e.g., earth dams) for which porewater pressure effects may be important.

Probabilistic Methods

Probabilistic methods of analysis differ from the deterministic methods described above in that they can account for the uncertainties inherent in representing seismic input motions, soil properties, etc., through the use of probabilistic models. Such methods can be based on either total stress or effective stress concepts, providing results in terms of a probability level associated with the occurrence of a given soil or structure response level (e.g., a given state of stress in the structure, the occurrence of liquefaction, etc.). As previously noted, probabilistic methods are of considerable interest among earthquake and geotechnical engineers and are a subject of extensive research; however, to be applicable to port and harbor facilities, such methods require further development of (a) practical engineering procedures that feature probabilistic representations of the soil/structure system response and the liquefaction potential of the site soil materials in accordance with Requirements 1 and 2; and (b) an extended experimental data base to enhance our ability to define probabilistic models of soil behavior.

Probabilistic approaches have been developed to a limited extent for soil/structure interaction analysis and to a somewhat greater degree for liquefaction assessment. The only readily available soil/structure interaction approach treats the seismic input motions probabilistically and the soil and structure properties deterministically [38]. A variety of probabilistic liquefaction procedures have been developed, including cumulative damage approaches [8] and methods based on seismic risk concepts [53, 54]. Only a very few current liquefaction analysis procedures feature a probabilistic definition of the soil properties [e.g., 14]. No procedures are yet available that treat both the soil/structure system response and the liquefaction potential in a probabilistic manner, although some suggestions for further development along these lines are provided in our research program report [51].

ILLUSTRATIVE DYNAMIC ANALYSIS

The final task of this research program consisted of a dynamic analysis of an actual sheet-pile bulkhead struture in Japan, carried out in order to illustrate the applicability of dynamic analysis procedures to port and harbor facilities. During the 1978 Miyagi-Ken-Oki earthquake, this structure suffered damage in the form of settlement and cracking of its apron.

The bulkhead structure that was considered is comprised of a steel sheet pile supported by equally spaced steel pipes and anchored by a tie-rod/H-beam system. An unreinforced concrete apron rests on the ground surface and extends from the sheet pile to beyond the anchor (Fig. 1). The following two sets of soil conditions were considered in the dynamic analysis of this structure: (a) the original soil conditions at the site, which consist of a layer of loose fill underlain by saturated sands and silt; and (b) modifications to these original soil conditions, as resulting from application of deep-compaction soil improvement techniques. Ground motions measured at a nearby site during the Miyagi-Ken-Oki earthquake were used as seismic input to the analysis.

The analytical procedure used in this illustrative example was a deterministic total stress method comprised of the following three steps: (a) a two-dimensional soil/structure interaction analysis carried out using the FLUSH code [22], which features an equivalent linear approach to represent the strain-dependence of the dynamic soil properties; (b) conversion of the irregular shear stress histories in the soil, as obtained from Step a, to equivalent cyclic stresses using the empirical approach described by Seed et al. [44]; and (c) assessment of the liquefaction potential at the site, through comparison of these cyclic shear stresses to cyclic stresses shown by the Seed et al. [45] field performance correlations to lead to liquefaction.*

The results obtained from this analysis are described in detail in our research program report [51]. The main trends from these results are summarized as follows:

---

*For this particular site, no laboratory cyclic test data were available which also would serve to assess the potential for liquefaction.

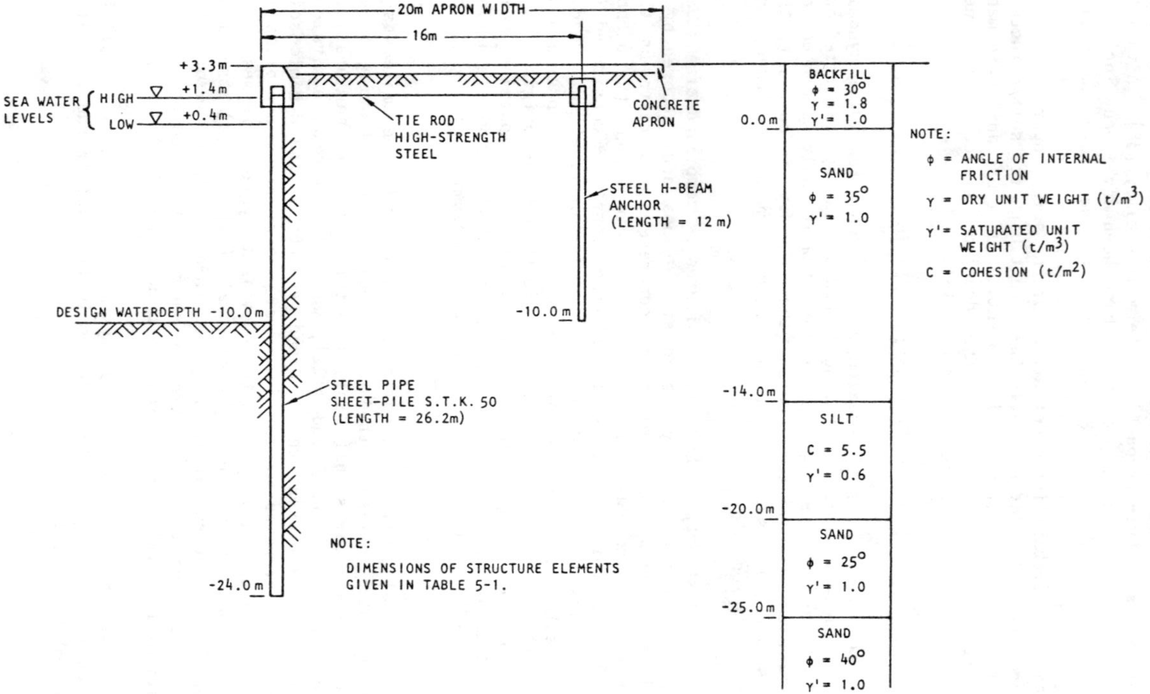

(a) Soil/structure configuration (b) Soil profile

FIGURE 1. SHEET-PILE BULKHEAD NO. 4 AT NAKANO WHARF, SENDAI HARBOR, JAPAN [48]

- The calculations indicated that the site of this sheet-pile bulkhead is susceptible to porewater pressure effects. These effects, together with possible earthquake-induced settlement of the unsaturated loose fill above the water table, provide a plausible explanation for the observed damage to this structure from the earthquake – settlement and cracking of the apron.

- Upper bound estimates of the earthquake-induced internal stresses in the structure indicated these structure stresses to be low when compared to anticipated design values, although the stresses in the apron are sufficient to have caused some cracking.

- Deep compaction soil improvement techniques were represented in this analysis by increasing the relative density of the near-surface soil layers in accordance with recommendations of Mitchell and Katti [26]. On this basis, such techniques were shown to substantially enhance the resistance of the upper portion of this site to porewater pressure buildup and liquefaction. This trend is in line with prior observations in Japan of the effects of soil improvement on the behavior of sand deposits during earthquakes [17, 35, 50].

- For these particular site conditions and seismic excitations, the dynamic analysis results produced lateral pressure distributions along the sheet-pile/soil interface that were markedly different from the pseudostatic lateral pressure distributions computed from the Mononobe-Okabe equation and commonly used for design.

- Comparisons of the two-dimensional soil/structure system response with one-dimensional free-field response results showed that the peak shear stresses in the near-surface regions of the soil backfill were affected by the topography of the site (and probably to a smaller degree by soil/structure interaction for this particular system). These effects on the soil/structure system response spectra and on the shear stresses in the deeper soil layers were much smaller.

CONCLUSIONS

The main findings of this research program are as follows:

- Lessons from Past Earthquakes. Prior earthquakes have shown that port and harbor facilities are susceptible to extensive damage from earthquakes, primarily due to porewater pressure buildup in the site soil materials. This damage has been particularly severe when large-scale liquefaction and/or massive submarine slides induced by porewater pressure buildup have occurred. Porewater pressure buildup has also led to extensive damage due to excessive lateral pressures applied to quay walls and bulkheads along the structure/backfill interface.

- Seismic Design Provisions. Seismic effects are often not nearly as important a design consideration for port and harbor facilities as would be desirable, in view of the extensive damage to such facilities that has occurred during prior earthquakes. The seismic design provisions that do exist typically address some of the potential earthquake-induced phenomena in a simplified pseudostatic manner and ignore many of the others.

- Dynamic Analysis Techniques. Dynamic analysis, which is now routinely used only for liquefaction assessments under free-field conditions, should be incorporated to a much greater extent into seismic design provisions for port and harbor facilities. Deterministic total stress methods are applicable at this time to the dynamic analysis of port and harbor facilities, including the soil/structure system response and the potential for significant porewater pressure buildup in the surrounding soil medium. The continuing development of deterministic effective stress methods and probabilistic methods for such applications should add significantly to these existing dynamic analysis capabilities.

ACKNOWLEDGEMENTS

This research program was carried out through a research grant to Agbabian Associates (AA) by the National Science Foundation (Grant No. CEE-8012337). This support is gratefully acknowledged. The authors wish to acknowledge the significant contributions of R.M. Pyke, geotechnical engineer from Berkeley California, who served as geotechnical consultant to the program. In addition, the program was enhanced by valuable discussions with H.B. Seed of the University of California at Berkeley, by extensive information provided to us by H. Tsuchida and S. Noda of the Port and Harbour Research Institute of Japan, and by personal visits and technical discussions with personnel from consulting engineering firms (Dames & Moore, Inc., Ertec Western, Inc.) from government agencies (Naval Civil Engineering Laboratory and the Naval Facilities Engineering Command), and from port authorities (Ports of Los Angeles, Long Beach, Oakland, and San Francisco). AA personnel who provided substantial support to this work were C.M. St. John (dynamic analysis evaluations), H.S. Ts'ao (analysis computations), P. LaPonza (typist), and G. Lillegraven and E. Harding (illustrators).

REFERENCES

1. Bathe, K.J., "ADINA, A Finite Program for Automatic Dynamic Incremental Nonlinear Analysis," Report 82448-1, Massachusetts Institute of Technology, Dec 1978.

2. Biot, M.A., "Theory of Elasticity and Consolidation for a Porous Isotropic Solid," Journal of Applied Physics, Vol. 12, 1941, pp. 155-164.

3. Booker, J.R., et al., "GADFLEA – A Computer Program for the Analysis of Pore Pressure Generation and Dissipation During Cyclic

or Earthquake Loading," EERC 76-24, Berkeley: Univ. of Calif., Earthquake Engineering Research Center, Oct 1976.

4. Chakrabarti, S., et al., "Seismic Design of Retaining Walls and Cellular Cofferdams," Proc. ASCE Geotechnical Engineering Division, Specialty Conf. on Earthquake Engineering and Structural Dynamics, Pasadena, CA, Jun 1978, pp. 325-341.

5. Chopra, A.K., "Hydrodynamic Pressures on Dams during Earthquakes," Journal of the Engineering Mechanics Division, ASCE, Vol. 93, No. EM6, Dec 1967, pp. 205-223.

6. Chopra, A.K., and Chakrabarti, P., "Earthquake Analysis of Gravity Dams Including Dam-Water-Foundation Interaction," Earthquake Engineering and Structural Dynamics, Vol. 9, No. 4, Jul-Aug 1981, pp. 363-383.

7. Departments of the Army, Navy, and Air Force, "Seismic Design for Buildings," TM 5-809-10/NAVFAC P-355/AFM 88-3, Chapter 13, Washington DC, Apr 1973.

8. Donovan, N.C., and Singh, S., "Liquefaction Criteria for Trans-Alaska Pipeline," Journal of the Geotechnical Engineering Division, ASCE, Vol. 104, No. GT4, Apr 1978, pp. 447-462.

9. Duke, C.M., and Leeds, D.J., "Response of Soils, Foundations, and Earth Structures to the Chilean Earthquakes of 1960," Bulletin of Seismological Society of America, Vol. 53, No. 2, Feb 1963, pp. 309-357.

10. Finn, W.D.L., "Liquefaction Potential: Developments Since 1976," Proc. International Conference on Recent Advances in Geotechnical Earthquake Engineering and Soil Dynamics, St. Louis, 1981, pp. 655-681.

11. Finn, W.D.L., and Martin, G.R., "Offshore Pile Foundations in Sand under Earthquake Loading," Applied Ocean Research, Vol. 2, No. 2, 1981, pp. 81-84.

12. Finn, W.D.L., et al., "An Effective Stress Model for Liquefaction," Journal of the Geotechnical Engineering Division, ASCE, Vol. 103, No. GT6, Jun 1977, pp. 517-533.

13. Ghaboussi, J., and Dikmen, S.U., "LASS-3, Computer Program for Seismic Response and Liquefaction of Layered Ground Under Multi-directional Shaking," UILU-ENG-79-2012, Urbana: Univ. of Ill., Jul 1979.

14. Haldar, A., and Tang, W.H., "Probabilistic Evaluation of Liquefaction Potential," Journal of the Geotechnical Engineering Division, ASCE, Vol. 105, No. GT2, Feb 1979, pp. 145-163.

15. Hall, J.F., and Chopra, A.K., "Two-Dimensional Dynamic Analysis of Concrete Gravity and Embankment Dams Including Hydrodynamic Effects," Earthquake Engineering and Structural Dynamics, Vol. 10, No. 2, Mar-Apr 1982, pp. 305-332.

16. Hayashi, S., et al., "Recent Revision of Design Standards on Seismic Effects for Port and Harbour Structures," Wind and Seismic Effects, Proc. Seventh Joint Conf. U.S.-Japan Cooperative Program in Natural Resources, Tokyo, May 1975.

17. Ishihara, K., et al., "Liquefaction Characteristics of Sand Deposits at an Oil Tank Site during the 1978 Miyagi-Ken-Oki Earthquake," Soils and Foundations (Japan), Vol. 20, No. 2, Jun 1980, pp. 97-111.

18. Japanese Port and Harbour Research Institute, Damage to Harbour Structures by the Niigata Earthquake, Part I (1964) and II (1965). (in Japanese)

19. Japanese Port and Harbour Research Institute, "Earthquake Resistant Design for Quaywalls and Piers in Japan," Earthquake Resistant Design for Civil Engineering Structures, Earth Structures and Foundations in Japan, compiled by Japan Society of Civil Engineering, 1980.

20. Katsikas, L.A., and Wylie, E.G., "Sand Liquefaction: Inelastic Effective Stress Model," Journal of the Geotechnical Engineering Division, ASCE, Vol. 108, No. GT1, Jan 1982, pp. 63-81.

21. Kitajima, S., and Uwabe, T., Analysis on Seismic Design in Anchored Sheet-Piling Bulkheads, Japanese Ports and Harbour Research Institute, Vol. 18, No. 1, Mar 1979.

22. Lysmer, J., et al., "FLUSH--A Computer Program for Approximate 3-D Analysis of Soil-Structure Interaction Problems," EERC 75-30, Berkeley: Univ. of Calif., Earthquake Engineering Research Center, Nov 1975.

23. Lysmer, J., et al., "SASSI: A System for Analysis of Soil-Structure in Interaction," UCB/GT/81-02, Berkeley: Univ. of Calif., Dept. of Civil Engineering, Apr 1981.

24. Margason, E., "Pile Bending during Earthquakes," Design, Construction, and Performance of Deep Foundations, Seminar Series, ASCE Geotechnical Group and Continuing Education Committee, San Francisco, 1975.

25. Martin, P.P., and Seed, H.B., "Simplified Procedure for Effective Stress Analysis of Ground Response," Journal of the Geotechnical Engineering Division, ASCE, Vol. 105, No. GT6, Jun 1979, pp. 739-758.

26. Mitchell, J.K., and Katti, R.K., "Soil Improvement: State-of-the-Art," Proc. Tenth International Conf. Soil Mechanics and Foundation Engineering, Stockholm, Jun 1981, Session 12.

27. Mononobe, N., "Earthquake-Proof Construction of Masonry Dams," Proc. World Engineering Conf., Vol. 9, 1929, p. 275.

28. National Academy of Sciences, The Great Alaska Earthquake of 1964, Engineering, Washington, DC, 1973.

29. Newmark, N.M., "Effects of Earthquakes on Dams and Embankments," Geotechnique, Vol. 15, No. 2, Jun 1965, pp. 139-160.

30. Noda, S., and Hayashi, S., "Damage to Port Structures by the 1978 Miyagi-Ken-Oki Earthquake," Proc. Seventh World Conf. on Earthquake Engineering, Istanbul, Turkey, Sep 1980, Vol. 9, pp. 415-423.

31. Noda, S., and Uwabe, T., "Seismic Disasters of Gravity Quay Walls," Technical Note No. 27, Japanese Port and Harbour Research Institute, 1975. (in Japanese)

32. Noda, S., and Uwabe, T., "Relation between Seismic Coefficient and Ground Acceleration for Gravity Quaywalls," Proc. Eighth US-Japan Conf. on Wind and Seismic Effects, Gaithesburg, MD, May 1976.

33. Okabe, S., "General Theory of Earth Pressure," Journal Japanese Society Civil Engineering, Vol. 12, No. 1, 1926.

34. Okamato, S., Introduction to Earthquake Engineering, John Wiley and Sons, New York, 1973.

35. Ohsaki, Y. "Effects of Sand Compaction on Liquefaction during the Tokachioki Earthquake," Soil and Foundations (Japan), Vol. 10, No. 2, Jun 1970, pp. 113-128.

36. Pita, F.W., et al., "Seismic Design of Seattle Waterfront Facilities," Journal of the Technical Councils of ASCE, Vol. 108, No. TC1, May 1982, pp. 24-33.

37. Pyke, R.M., et al., "Liquefaction Potential of Hydraulic Fills," Journal of the Geotechnical Engineering Division, ASCE, Vol. 104, No. GT11, Nov 1978, pp. 1335-1354.

38. Romo-Organista, M.P., et al., "PLUSH--A Computer Program for Probabilistic Finite Element Analysis of Seismic Soil/Structure Interaction," EERC-77/01, Berkeley: Univ. of Calif., Earthquake Engineering Research Center, Sep 1980.

39. Schnabel, P.B., et al., "SHAKE--A Computer Program for Earthquake Response Analysis of Horizontally Layered Sites," EERC-72-12, Berkeley: Univ. of Calif., Earthquake Engineering Research Center, Dec 1972.

40. Scott, R.F., "Earthquake-Induced Earth Pressure on Retaining Walls," Proc. Fifth World Conf. on Earthquake Engineering, Rome, Jun 1973, Paper 202.

41. Seed, H.B., "Soil Liquefaction and Cyclic Mobility Evaluation for Level Ground during Earthquakes," Journal of the Geotechnical Engineering Division, ASCE, Vol. 105, No. GT2, Feb 1979, pp. 201-255.

42. Seed, H.B., "Considerations in the Earthquake-Resistant Design of Earth and Rockfill Dams," Geotechnique, Vol. 29, No. 3, 1979, pp. 215-263.

43. Seed, H.B., and Whitman, R.V., "Design of Earth Retaining Structures for Dynamic Loads," Proc. ASCE 1970 Specialty Conf. on Lateral Stresses in the Ground and Design of Earth-Retaining Structures, Jun 1970, pp. 103-147.

44. Seed, H.B., et al., "Representation of Irregular Stress Time Histories by Equivalent Uniform Stress Series in Liquefaction Analyses," EERC 75-29, Berkeley: Univ. of Calif., Earthquake Engineering Research Center, Oct 1975.

45. Seed, H.B. et al., "Evaluation of Liquefaction Potential Using Field Performance Data," Journal of the Geotechnical Engineering Division, ASCE, Vol. 109, No. GT3, Mar 1983, pp. 458-482.

46. Taylor, P.W., and Indrawan, Z., "A Simple Method of Estimating Seismic Pressure from Cohesive Soils against Basement Walls," Proc. International Conf. on Recent Advances in Soil Dynamics and Earthquake Engineering, St. Louis, Apr 1981, pp. 241-244.

47. Terzaghi, K., Die Berechnung der Durchlassigkeitsziffer des Tones aus dem Verlauf der Hydrodynamischen Spannungserscheinungen, Sitznurngber. Akad. Wiss, Wien, 132, 125, 1923.

48. Tsuchida, H., et al., "The Damage to Port Structures by the 1978 Miyagi-Ken-Oki Earthquake," Technical Note No. 325, Japanese Port and Harbour Research Institute, 1979. (in Japanese)

49. U.S. Navy Facilities Engineering Command, "Design Manuals," DM-7, DM-25, and DM-26, Washington DC, 1968-71.

50. Wanatabe, T., "Damage to Oil Refinery Plants and a Building on Compacted Ground by the Niigata Earthquake and their Restoration," Soil and Foundations (Japan), Vol. 6, No. 2, Mar 1966, pp. 86-99.

51. Werner, S.D., and Hung, S.J. "Seismic Response of Port and Harbor Facilities," R-8122-5395, El Segundo, CA: Agbabian Associates, Oct 1982.

52. Westergaard, H.M., "Water Pressure on Dams during Earthquakes," Transactions American Society of Civil Engineers, Vol. 98, Paper No. 1835, 1933, pp. 419-433.

53. Yegian, M.K., and Whitman, R.V., "Risk Analysis for Ground Failure by Liquefaction," Journal of the Geotechnical Engineering Division, ASCE, Vol. 104, No. GT7, Jul 1978, pp. 921-938.

54. Youd, T.L., and Perkins, D.M., "Mapping Liquefaction-Induced Ground Failure Potential," Journal of the Geotechnical Engineering Division, ASCE, Vol. 104, No. GT4, Apr 1978, pp. 433-446.

55. Zienkiewicz, O.C., et al., "Effective Stress Dynamic Modelling for Soil Structures Including Drainage and Liquefaction," Proc. International Symposium on Soils under Cyclic and Transitive Loading, Swansea, England, Jan 1980, pp. 551-554.

Performance of Water and Sewer Lifelines in the May 2, 1983 Coalinga, California Earthquake

Jeremy Isenberg[1] MASCE and Craig E. Taylor[2] MASCE

ABSTRACT

Water supply and filtration systems for the City of Coalinga suffered little damage due to the May 2, 1983 earthquake. The water transmission system suffered five breaks. The urban distribution system was disrupted, but the number of breaks and leaks reported immediately following the earthquake may be smaller than would be predicted based on past performance of other systems. Leak rates in the water distribution system remained higher than normal for Coalinga for several months following the May 2 earthquake. There were three breaks in the sanitary sewer system, and numerous blockages whose removal continues to add to the burden of post-earthquake recovery.

INTRODUCTION

Water supply and filtration systems for the City of Coalinga sustained little damage due to the May 2, 1983 earthquake. The transmission system suffered breaks or leaks in five known locations. The urban distribution system suffered leaks and breaks which left users locally without service in some cases up to several days; as has been noted in previous earthquakes in California and Washington state, the earthquake and its aftershocks have at least temporarily accelerated leak rates to levels higher than normal for this system. Some wells which provide irrigation for agricultural land sustained significant damage.

Three breaks occurred in the sanitary sewer system. There were numerous blockages whose removal added to the burden of post-earthquake recovery. No significant damage to the sewer treatment plant occurred.

DESCRIPTION OF DAMAGE

City of Coalinga--Immediate Damage

Water for the City of Coalinga is supplied by the State Water Project through a series of smaller canals to the City of Coalinga Water Filtration Plant (see Fig. 1). After treatment, water is pumped through a 27-inch cement-lined pipeline with bell-and-spigot joints to Palmer

[1]Weidlinger Associates, 3000 Sand Hill Rd., Bldg. 4, Suite 155, Menlo Park, CA 94025.
[2]J. H. Wiggins Company, 1650 So. Pacific Coast Hwy., Redondo Beach, CA 90277-5674.

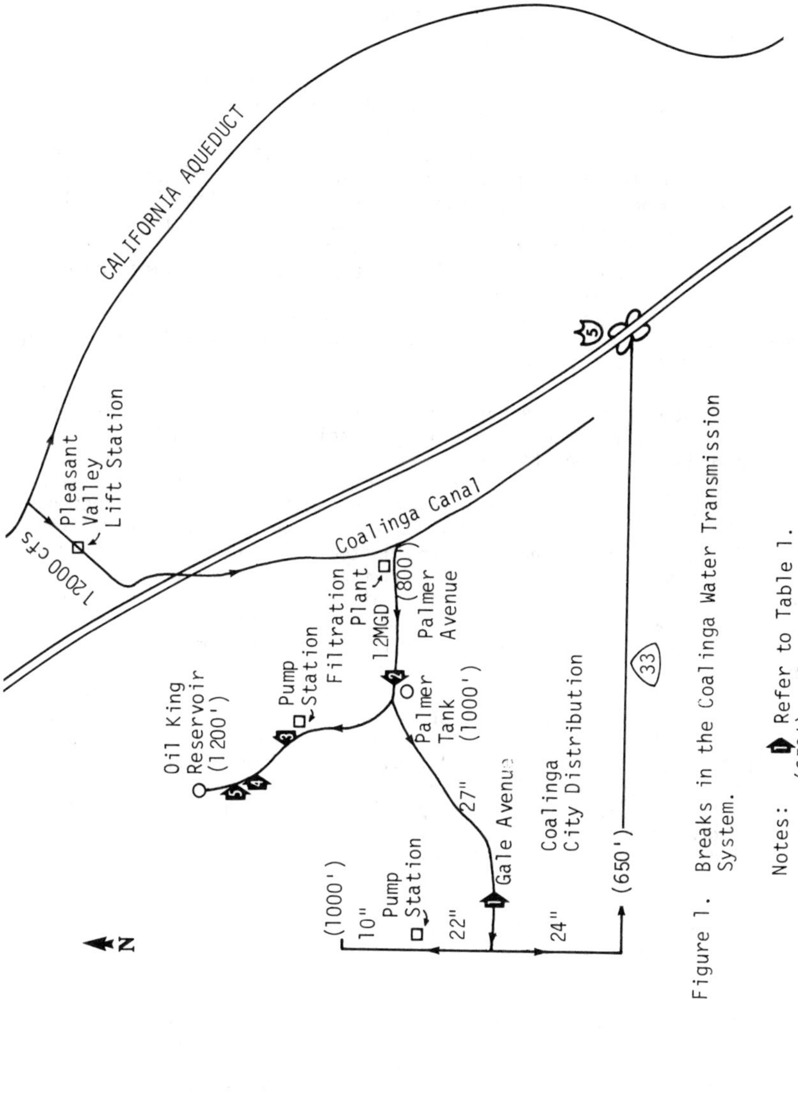

Figure 1. Breaks in the Coalinga Water Transmission System.

Notes: 1 Refer to Table 1.
(650') refers to elevation.

storage tank. Water then flows by gravity through the same size and type of transmission line to the city distribution system. A 10-inch cement-lined steel pipe with rubber gasket bell-and-spigot joints serves oil fields north of the city; this leg of the system includes a pump station and Oil King storage tank.

No damage was found in reaches of the San Lucas, Pleasant Valley and Coalinga Canals (1). Cracks occurred in soil adjacent to the San Lucas Canal; the cracks were overlain by "mole tracks" parallel to the canal in a zone 8-12 inches wide and 4-8 inches high. Gaps at the ends of a bridge abutment at Pump Station 17R were 2 inches on the east and 1/2 inch on the west. At the Pleasant Valley Lift Station, one outlet structure cracked jamming the sluice gate in an open position. Numerous cracks, up to 3/16-inch wide, appear in the parking area surrounding the outlet. A surge tank between the Pleasant Valley Lift Station and outlet lifted 1 1/2-2 inches; the tank had not yet been inspected for damage other than stretched bolts. Numerous overhead lights fell to the floor of the Pleasant Valley Lift Station; transformers shifted on their foundations 1/2 inch or less.

No significant damage was found at the City of Coalinga Water Filtration Plant. Electrical panels did not shift, even though they were not bolted down. Two bolts sheared at the top of gussets in the diagonal bracing system of the Chemical Fuel Building frame. A pre-existing crack in the water clarifier was inspected and found to be unaffected by the earthquake. At the anchored washwater standpipe tank, circumferential welds were damaged near the tank bottom. A leak in the discharge connection was repaired easily. In addition, a flexible coupling failed between two mismatched discharge pipes at the pump station. However, both the tank and the pump station remained operational in spite of the damage. There is evidence of minor shifting of other equipment, but none for the pumps and no damage to them.

Two leaks in the 27-inch transmission line were found (Fig. 1). One on Palmer Avenue near the holding tank occurred in the threaded connection to a 2-inch steel pressure equilization line which is connected to the crown of the transmission line. By the morning of Wednesday, May 4, a hole appeared in the surface of Palmer Avenue. No evidence of adjacent ground rupture was found. A larger leak occurred on Gale Avenue west of the intersection with California 33/198 due to tension failure at a joint. Roughly N-S trending cracks appeared in the plowed field south of Gale. The largest of these was 150 yards west of the break; other cracks appeared adjacent to the break. There was flooding of the verge and adjacent field. The rupture was repaired by replacing a 10-ft section.

In addition, three breaks occurred on the 10-inch line serving the oil fields. All of these were compression failures at joints. Near the pump station, the failure may be related to a vertical offset which appeared in the surface of the adjacent highway to be about 1 1/2 inches. The other two occurred at points where the vertical profile changes rapidly and where the pipe is in roadway fill. Cracks several feet deep appeared near the crest of the fill embankment covering the pipe and trending roughly parallel to them. In the substation that serves the pumps for this line, electrical equipment struck the wall and an outside wire brace fell; service was not interrupted. At the Oil King tank, a cast iron coupling failed. A summary is given in Table 1.

The distribution system in the City of Coalinga includes approximately 27 miles of pipe whose materials are indicated in Table 2. Location of known breaks are shown in Fig. 2. The breaks and leaks for which more detailed information is available are summarized in Table 3. Numerous other repairs were required to 4-inch and 6-inch wrought or cast iron pipes. Many of these were made by crews from outside the Coalinga Department of Public Works, for which documentation is unavailable. Some old corporation stops for disused service connections opened, possibly due to surges or to improper plugging or both. Many of these breaks and leaks occurred in the barrel. The ages of wrought iron (WI) and cast iron (CI) pipes which leaked or broke cannot be estimated accurately, but the impression is that they are among the older pipes in the system, including some from about 1900. Many sections were corroded. For example, the repair crew on the Fresno Avenue leak at Washington said that reach of pipe had been mended frequently due to corrosion-induced leakage. The approximate locations of these leaks are shown in Fig. 2. Also shown are reaches of pipe deemed too old or badly damaged to justify point repairs; in these cases, whole reaches of pipe were replaced.

Overall water distribution pipe performance is summarized in Table 4, which contains only known breaks and serious leaks, especially those recalled by utility personnel or else reported in damage survey reports. In addition, single reaches of pipe where damage occurred were counted only once in Table 3. Based on Table 3 and also data on other earthquakes, it can be speculated that the Coalinga water distribution system performed comparatively poorly. The shaking intensity in Coalinga itself has been estimated as VIII (on the Modified Mercalli Scale), and no ground failures are known to have affected distribution pipes [8]. Accordingly, one may compare Coalinga pipe damage data with expected damage rates based on previous earthquake data. Figure 3 provides one such graphic comparison. Overall, cast iron (CI) pipe in the Coalinga system performed slightly worse than expected (relative to past damage rates) and asbestos-cement (AC) pipe in the Coalinga system performed slightly worse than expected relative to past damage rates. Aggregate shaking damage in the Coalinga water distribution system was higher than expected, given the severity of shaking, possibly due to pre-existent piping system weaknesses (corrosion, high normal leak rates, older pipe, etc.).

City of Coalinga--Ongoing Effects

In four U.S. earthquakes (1965 Puget Sound (Seattle); 1969 Santa Rosa; 1971 San Fernando; and 1979 Imperial County), (2,3) corrosion, especially in ferrous pipes, has been identified as a factor which increases the number of leaks. In addition, recordkeeping by the Los Angeles Department of Water and Power is sufficiently detailed to show that the 1971 San Fernando earthquake accelerated leak rates for several years after the event. This is attributed to the action of ground shaking and pressure surges toward weakening pipes already made locally thin by corrosion. Whereas a sustained corrosion control program by the LADWP reduced the rate of leakage over a period of about 15 years prior

Table 1. Damage to water transmission lines in the 2 May 1983 Coalinga earthquake*.

Reference No.	Location	Pipe Material	Pipe Diameter	Comments
1	Gale Ave. 500' west of Hwy. 33	Steel-lined concrete	27"	Tension failure at bell-and-spigot joint. Fish mouth developed after surges. Near lateral ground fissure.
2	Palmer Ave. 300' east of Palmer tank	"	27"	Near Coalinga anticline. Failure in 2" threshold steel coupling.
3	East side of Hwy. 33 between pump station and Hwy. 198	"	10"	Compression failure. Located near road pavement slumping. Concrete chipped away. Road split 1-1 1/2" vertically nearby.
4	Northeast of Hwys. 198/33 junction, between Oil King Reservoir and pump station	"	10"	Compression failure. Located near voids in road fill due to likely vertical offsets exceeding 12". Concrete chipped away. 2" flexibility.
5	Same as 4	"	10"	Same as 4. Failure 40' apart (single reach). Ruptured joints.

*All constructed in 1972.
Note: See Figure 1.

Table 2. Summary estimate of pipe in place in Coalinga Water Distribution Network, 1983.

	Subdivision Zone Number and Approximate Age				
	1 (Pre 1900)	2 (1900-1917)	3 (1920-1941)	4 (1945-1961)	5 (1961-1983)
Asbestos-Cement (68,480')					
Concrete/Steel (10,530')					
Cast Iron (45,420')					
Welded Iron (18,140')					
Total (ft)	25,150	37,070	21,770	38,690	19,270

Notes: 1. Age of pipes may be less than that of subdivision.
2. Small amount of PVC is excluded.

10,000 ft

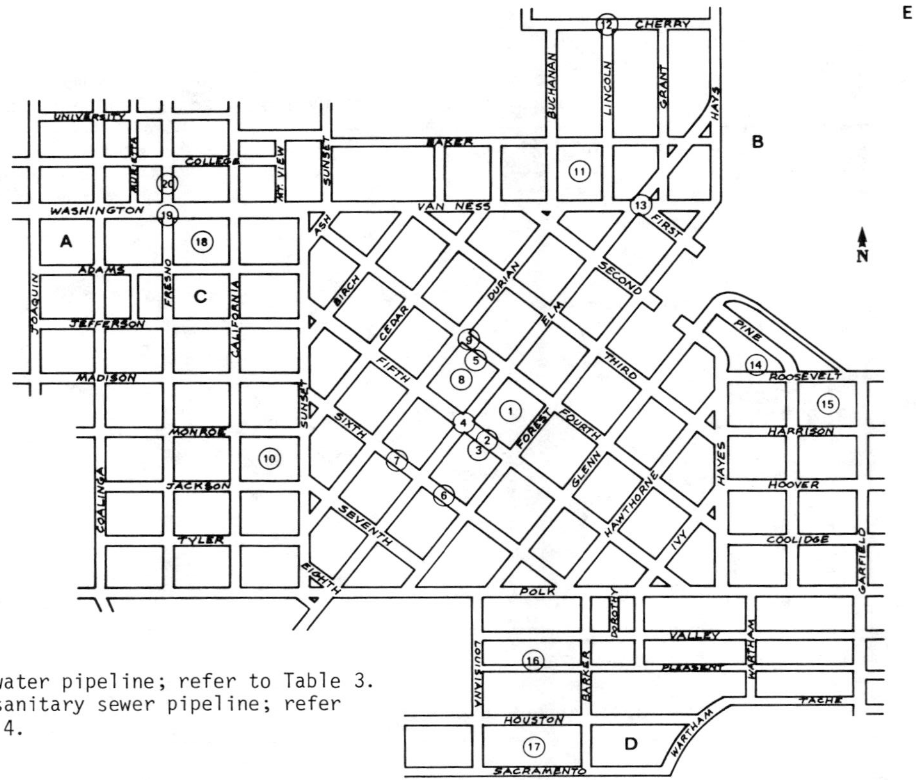

○ - Damaged water pipeline; refer to Table 3.
D - Damaged sanitary sewer pipeline; refer to Table 4.

Figure 2. Map of central Coalinga showing locations of damage to water and sanitary sewer pipelines.

Table 3. Damage to water distribution lines in 2 May 1983 Coalinga earthquake.

Ref. No.	Location	Pipe Material	Pipe Diameter	Previous Poor Performance[1]	Likely Corrosion Related	Comments
1	Elm/Forest Alley bet. Fourth & Fifth	AC	8"			Snapped at bell-and-spigot joint
2	Fifth bet. Elm & Forest	AC	8"	Yes		Barrel crack; 3 valves replaced
3	Fifth bet. Elm & Forest	AC	4"	Yes		Service sprinkler snapped at bell-and-spigot joint
4	Elm & Fifth	AC	8"			Snapped at bell-and-spigot joint
5	Elm/Durian off Fourth	AC	6"	Yes	Yes	Broke both at joint and in barrel; highly corrosive area
6	Elm/Forest on Sixth	AC	8"			
7	Elm/Durian on Sixth	CI	10"		Yes	Highly corrosive area
8	Elm/Durian bet. Fourth & Fifth	CI	4"		Yes	Highly corrosive area
9	Elm/Durian on Fourth	CI	6"		Yes	4 valves put in; DSR request turned down
10	Monroe/Jackson bet. California & Sunset	CI	6"	Yes		Snapped at bell-and-spigot joint 2" from bell
11	Buchanan/Lincoln bet. Van Ness & Baker	CI	6"			Repaired with clamp
12	Cherry & Lincoln	CI	6"			
13	Van Ness bet. Lincoln & Grant	CI	6"			2 leaks, one repaired with clamp

Table 3. Concluded.

Ref. No.	Location	Pipe Material	Pipe Diameter	Previous Poor Performance[1]	Likely Corrosion Related	Comments
14	Roosevelt bet. Pine & Hayes	AC	6"			
15	Roosevelt/Harrison bet. Pine & Garfield	CI	4"			Damaged in 3 places close to same hole; damage brought out in aftershock. 2850 ℓf AC water main proposed
16	Pleasant bet. Louisiana & Barker	CI	4"	Yes	Yes	Corrosion leak in barrel; 2 leaks repaired with clamps
17	Houston/Sacramento bet. Louisiana & Barker	WI	4"	Yes		1 leak repaired with clamp
18	Washington/Adams bet. California & Fresno	WI	4"			1350 ℓf of 6" AC proposed; 60 ℓf replaced
19	Fresno & Washington	CI	4"			Split in barrel at 4 places
20	Fresno bet. College & Washington	CI	4"		Yes	Snapped at joint; 2 leaks in corroded pipe repaired with clamps

Notes: [1]According to Department of Public Works.
See Figure 2.
WI = Wrought Iron
CI = Cast Iron
AC = Asbestos-Cement

Table 4. Summary of pipe breaks, lengths (1,000 ft) and failure ratios for underground water distribution pipes in the 2 May 1983 Coalinga earthquake.

Pipe Material	Parameter	Pipe Diameter (inches) LT 6	8-11	12-16	18-30	Total
Asbestos-Cement	Breaks	3	4	0	0	7
	Length	26.53	41.32	0.69	0	68.48
	Ratio	0.11	0.10	0	0	0.10
Cast Iron	Breaks	10	1	0	0	11
	Length	37.77	7.65	0	0	45.42
	Ratio	0.26	0.13	0	0	0.24
Wrought Iron	Breaks	2	0	0	0	2
	Length	16.06	2.08	0	0	18.14
	Ratio	0.12	0	0	0	0.11
Polyvinyl Chloride (PVC)	Breaks	0	0	0	0	0
	Length	1.08	0	0	0	1.08
	Ratio	0	0	0	0	0
Concrete/Steel	Breaks	0	0	0	0	0
	Length	0	0	6.10	4.43	10.53
	Ratio	0	0	0	0	0

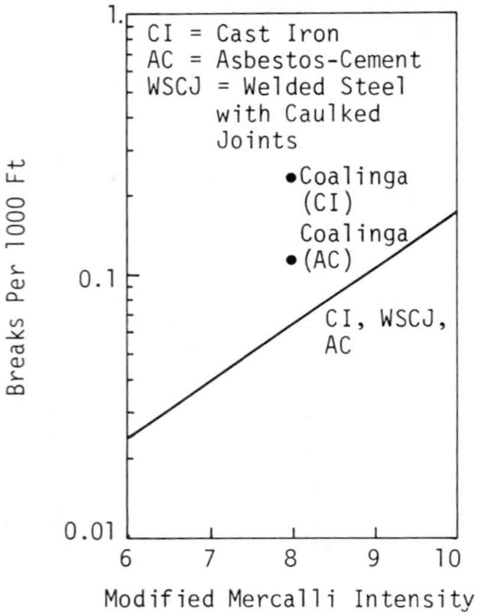

Figure 3. Coalinga water distribution pipe damage compared with previous earthquake vulnerability relationships for underground pipe in ground shaking area (sources: Table 3 and Ref. 7).

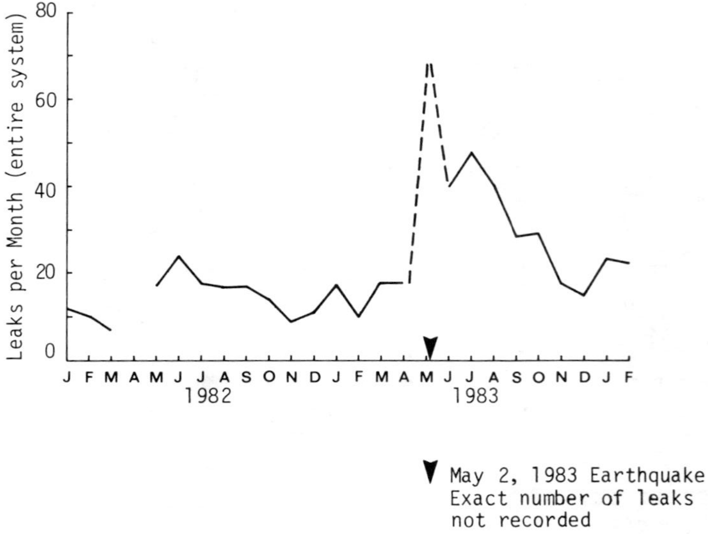

Figure 4. Leaks per month in Coalinga, California distribution system (approximately 27 miles).

to the 1971 earthquake, the trend was temporarily reversed for several years following it--at least in selected areas.

The Coalinga Department of Public Works does not have a formal corrosion control program. Resistivity of the soil ranges from 1,500 ohms/cm^3 to 10,000 ohms/cm^3, depending on moisture (4). A cathodic protection system has been in place for natural gas lines since 1978. This and other sources of stray current combined with moderate soil resistivity, at least in some seasons, and some aging ferrous pipes leads to the expectation that corrosion-related leakage will occur. This is consistent with Fig. 4, which shows the total number of leaks in each month for 16 months preceding the earthquake and 9 months following it. Although lingering effects of earthquake-related damage are still being investigated, it may be possible eventually to state how much the earthquake has shortened the life of the system.

Pleasant Valley Water District

The Pleasant Valley Water District comprises 39,000 acres east of Coalinga. Damage to wells, distribution pipelines and related electrical equipment within the District has been described by Summers (5). He has identified approximately 23 damaged wells from the May 2, 1983 earthquake and its aftershocks. Among these, damage to pumps was prevalent. For example, pump head misalignment was especially common. These pumps are at depths of 600 ft and greater. Instances of collapsed casings, diminished yield and damaged column assembly were also cited.

In addition, pipelines used to convey well water to fields for irrigation failed at joints and by collapse of vertical structures such as pump stands, air vents and stand pipes. Most farmers used surface pipes and other temporary means to meet summer irrigation needs. Long-term effects on the aquifer system cannot yet be assessed.

City of Coalinga Sanitary Sewage System

A 480-ft section of old concrete sewer pipe in the 300 block of Adams-Jefferson Alley west of downtown Coalinga was partially collapsed (see Fig. 2). In addition, a 40-ft section of 8-inch concrete interceptor parallel to Baker east of Hayes collapsed. Also, a joint failed in tension at an inverted syphon at Warthan Creek crossing (see Table 5). In addition to immediate damage, there are lingering effects in the form of increased frequency of blockages in older sections. This is believed due in part to damage sustained by grouted joints and to resulting vulnerability to root invasion.

The town sewage disposal plant, whose improvement and expansion construction work was almost complete at the time of the earthquake, did not sustain significant damage. The earthen dikes of several stabilizing ponds were disturbed and subsequent investigation may reveal minor damage. The plant continued to operate during and after the earthquake.

SUMMARY AND CONCLUSIONS

Main water supply to the City of Coalinga was not interrupted. However, the only transmission line suffered a leak, possibly due to

Table 5. Damage to sewer lines in the 2 May 1983 Coalinga Earthquake

Reference No.	Location	Pipe Material	Pipe Diameter	Comments
A	Adams-Washington & Joaquin-Coalinga Alleys	concrete	?	480-ft section caved in
B	Baker St. east of Hayes	concrete	8"	Caved in
C	Adams-Jefferson & Fresno-California Alleys	concrete	?	480-ft section partially collapsed; offset at middle of joint
D	Houston-Sacramento & Baker-Sacramento Alleys	?	6"	Snapped at bell-and-spigot joint
E	Warthan Creek Crossing			Interceptor line damaged at creek crossing

Note: See Figure 2.

abrupt local ground movement (i.e., surface crack); disruption could have been more serious due to lack of redundancy in the transmission system. Lingering effects of damage to water distribution and service lines and to sewer lines are observed. Apparently, these are inevitable where there is a significant corrosion problem. It is a matter for discussion whether costs of correcting lingering defects should be included in emergency recovery funds or should be treated as ordinary maintenance. Pumps suffered misalignment; possibly the damage is related to the relatively great depths of the pumps (600 ft and greater).

ACKNOWLEDGMENTS

The authors are indebted to Alan Jacobsen (Director of Public Works), Tito Balling and Ray Buchanan of the Department of Public Works, City of Coalinga, for their cooperation and assistance. Partial support for this research was furnished to J. Isenberg of Weidlinger Associates through NSF Grant CEE-8116842 and by the Earthquake Engineering Research Institute, which supported post-earthquake reconnaissance, and to C. Taylor, J.H. Wiggins Co., through NSF Grant CEE-82118677.

REFERENCES

1. Isenberg, J. and L. E. Escalante, "Damage to Lifelines," Coalinga, California, Earthquake of May 2, 1983. Reconnaissance Report, Earthquake Engineering Research Institute, Report No. 84-03, Jan. 1984.

2. Isenberg, J., "Role of Corrosion in Water Pipeline Performance in Three U.S. Earthquakes," U.S. National Conference on Earthquake Engineering, Stanford University, August 22-24, 1979.

3. Brandow, G. E., Coordinator, D. J. Leeds, Editor, "Reconnaissance Report, Imperial County, California, Earthquake, October 15, 1979," Earthquake Engineering Research Institute, February 1980.

4. Heath Consultants, Inc., "Corrosion Evaluation Survey, City of Coalinga, Coalinga, California," Report to the City of Coalinga, CE 3249, 1981.

5. Summers, J. B., "Damage to Irrigation Wells and Other Facilities in the Pleasant Valley Water District Due to the May 2, 1983 Earthquake and Aftershocks," California Department of Conservation, Division of Mines and Geology, Special Publication 66, 1984.

6. Borcherdt, R., "The Coalinga Earthquake Sequence Commencing May 2, 1983," U.S. Geological Survey, Open File Report 83-511, 1983.

7. Eguchi, R. T., M. R. Legg, C. E. Taylor, L. L. Philipson and J. H. Wiggins, "Earthquake Performance of Water and Natural Gas Supply Systems: Final Report," J. H. Wiggins Co., Redondo Beach, CA, July 1983.

8. Hopper, Margaret G., P. C. Thenhaus, Lynn Barnhard and S. T. Algermisser, "Damage Survey in Coalinga, California, for the Earthquake of May 2, 1983," California Division of Mines and Geology, Special Publication 66, 1984.

DYNAMIC RESPONSE CORRELATION OF CYLINDRICAL TANKS

George C. Manos* and Ray W. Clough**

The May 1983 Coalinga Earthquake provided an unique opportunity to correlate the actual earthquake performance of cylindrical liquid storage tanks with the behavior observed in laboratory shaking table experiments. A 38.5 ft diameter steel water tank which developed typical "elephant's foot" buckling has been found to be a good prototype for a 12 ft diameter aluminum tank that has been tested extensively with the University of California earthquake simulator. In this paper the properties of the prototype and model are correlated, as are the characteristics of the base motions applied in the field and in the laboratory. From these relationships, the measured dynamic stress response of the model is extrapolated to obtain an estimate of the stresses induced in the prototype by the Coalinga earthquake. The results clearly demonstrated that buckling was to be expected in this case.

INTRODUCTION

The magnitude M = 6.5 earthquake that occurred near Coalinga, California, on 2 May 1983 produced peak ground accelerations severe enough to cause a wide variety of structural damage. Because this is an oil production region, a large number of tanks for storage of oil and other liquids were subjected to intense shaking, and their behavior during the earthquake provided important information on the earthquake resistant design of ground supported cylindrical liquid storage tanks. The lead author of this paper visited the epicentral region several times after the earthquake and made observations on the performance of many tanks. One of these tanks has been chosen for discussion in this paper -- a relatively small steel water storage tank designated T400 that was part of a water treatment plant; the water was used in the oil production processes.

One reason this tank was selected for detailed evaluation is that it suffered very distinct "elephant's foot" buckling near the base; an equally important factor, however, is that it is a close equivalent of the full scale prototype of a tank model that has been studied extensively at the Earthquake Engineering Research Center (EERC) of the University of California. This 12 ft diameter aluminum model has been subjected to a wide range of simulated earthquake motions, using the 20 ft square EERC earthquake simulator facility. The laboratory investigation extended over a period of nearly seven years, and has been documented in a number of papers and reports [1,2,3,4,5] which give complete descriptions of the test procedures, instrumentation and

* Lecturer, Laboratory of Structural Analysis, Aristotelian University of Thessaloniki, Greece.

**Nishkian Professor of Structural Engineering, University of California, Berkeley, U.S.A.

results. It is the purpose of this paper to compare the laboratory performance of the test model with the observed behavior of prototype tank T400 as a result of the Coalinga earthquake. The performance of a number of other tanks damaged during the earthquake is described in another paper [6].

In order to correlate the behavior of the tank in the field with the measured shaking table response of the test model, it is necessary to compare: (a) the mechanical properties of the tanks, (b) the dynamic characteristics of the base motions to which they were subjected, and (c) the stress levels induced during the dynamic response. In this paper, these three types of comparison are discussed in sequence; conclusions are then drawn concerning the degree of damage observed in tank T400, and also with regard to current tank design procedures.

COMPARISON OF TANK PROPERTIES

Prototype Tank T400

Water storage tank T400 shown in Figs. 1 and 2, is 38.5 ft in diameter, 24 ft high and contained 18 ft depth of water at the time of the earthquake. It was formed of three courses of steel plate, with both horizontal and vertical joints connected by high tensile strength steel bolts. The thickness of the bottom course, in which the elephant's foot buckling occurred is 0.1793 in. As shown in Fig. 2, the top of the tank was closed by a conical roof.

The cross-section view of the tank site, Fig. 3, shows that a series of tanks had been excavated into a hill side in steps, and that T400 was located near the top of the hill. Because of this excavation, the foundation for the tank was a level hard soil surface; a 4 in. layer of gravel was placed directly under the tank bottom plate and extended one foot beyond the tank wall, where it was confined by a steel ring (Figs. 1 and 2). No anchoring devices were provided at the tank wall, so the tank was free to uplift from the foundation.

Laboratory Model Tank

The dimensions and configuration of the tank tested in the laboratory are shown in Fig. 4. It was formed of aluminum plating, and was originally designed as a one-third scale model of a typical (but hypothetical) 36 ft diameter steel prototype. However, in this study it is viewed as a 1:3.2 scale model of tank T400. Table 1 lists the dimensions and other properties of the prototype and of an exact 1:3.2 model, as well as the corresponding values for the laboratory test tank. This table shows that the most significant modelling discrepancies are in the tank height, and in the thickness of the lowest shell course and the bottom sketch plate. Laboratory studies have shown, however, that the depth of water in the tank is much more significant that the height of the tank wall, so the height discrepancy will not be considered further. Allowances are made for the discrepancies in water depth and in plate thickness as part of the procedure for correlating the behavior of model and prototype, as is discussed later. More comprehensive discussions of the model similitude requirements are presented in References 1 and 7.

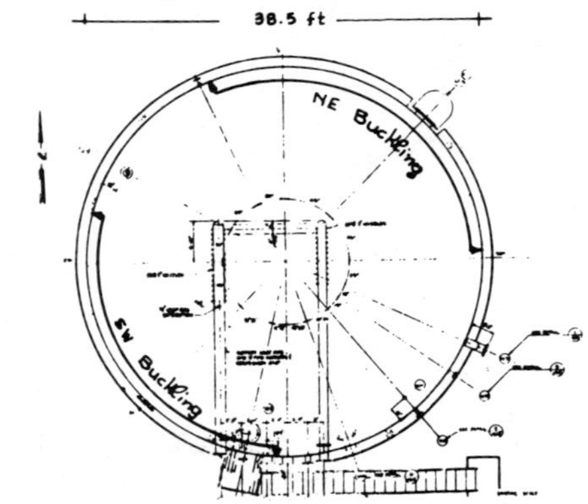

Fig. 1. Plan of tank T400

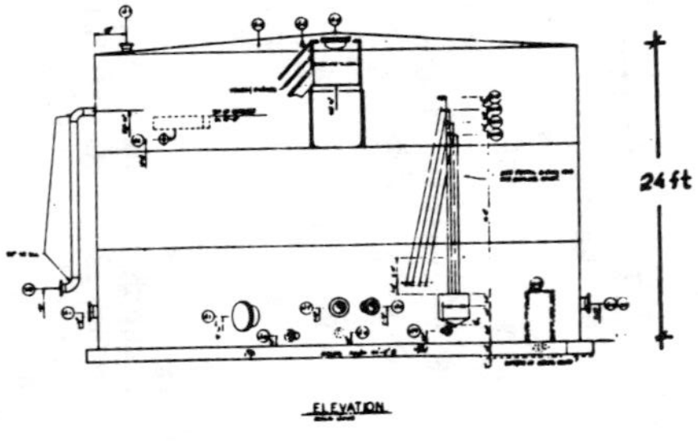

Fig. 2. Elevation of tank T400.

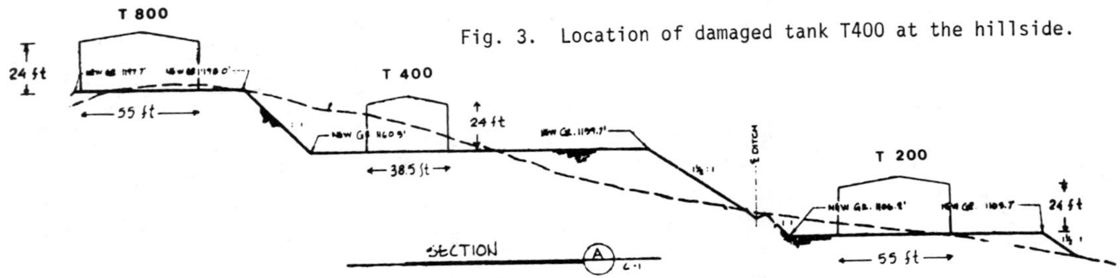

Fig. 3. Location of damaged tank T400 at the hillside.

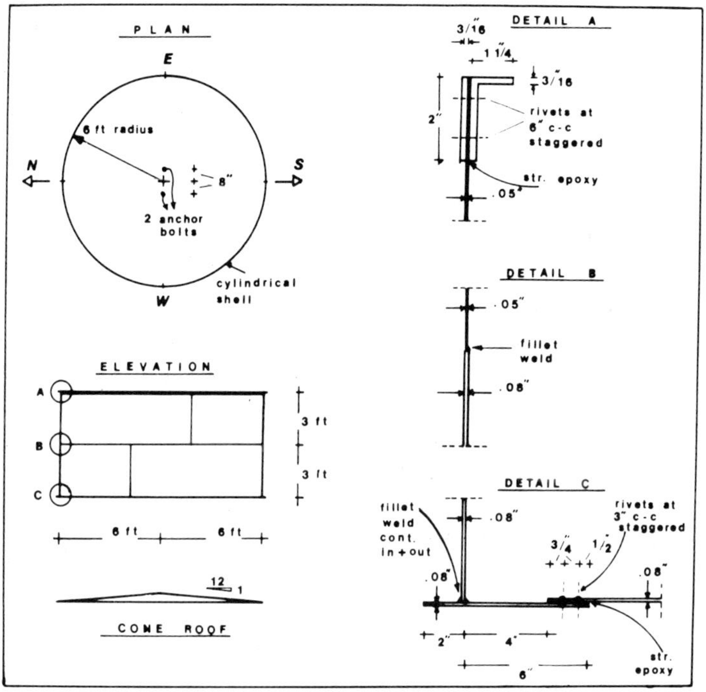

Fig. 4. Dimensions and structural details of model tank tested in the laboratory

Table 1: Pertinent Tank Properties

	Prototype	1/3.2 Scale Model	Actual Model
Diameter (ft)	38.5	12.03	12.00
Height (ft)	24.0	7.50	6.00
Water Depth (ft)	18.0	5.63	5.00
Shell Thickness at Bottom (in)	0.180	0.056	0.080
Bottom Sketch Plate Thickness (in)	0.134	0.042	0.080
Modulus of Elasticity (ksi)	30,000	9,375	10,000
Relative Vibration Period	1	1/1.79	1/1.73

Correlation of Prototype and Model Properties

Additional aspects of modelling that must be considered in comparing the model and prototype performance are as follows:

Connections of Tank Plating: The aluminum tank model had welded joints both in the shell plate connections and in the connection of side walls to the bottom, whereas the prototype steel tank had bolted joints as mentioned earlier. This modeling discrepancy would have been significant if the prototype or model had failed at the joints. However, the joint is considered to have had little effect on the tank deformations, including the elephant's foot bulge in the bottom course, so this discrepancy will be disregarded in the discussion that follows.

Tank Foundation: The prototype tank was free to uplift from a very firm, but not rigid, base consisting of a gravel layer over hard soil. The model tank was tested with three base conditions: (a) clamped against uplift from a rigid concrete base, (b) free to uplift from the rigid base, and (c) free to uplift from a sponge rubber pad interposed between the tank bottom and the concrete base (where the sponge rubber pad was chosen to simulate a very soft soil foundation, as described in Ref. 8). In the present comparison, it is assumed that the prototype support condition is between cases b and c above, but is much more similar to the rigid concrete base than to the sponge rubber pad.

Top Condition: The aluminum model was tested both with open top and with a conical roof; the model roof plating was far thicker than indicated by the scale ratio of 1:3.2 (0.150 in actual model vs 0.042 in scaled model). However, the main influence of the roof is to constrain the out-of-round distortion at the top of the tank shell, and it is believed that this constraint would be essentially complete even for the true scale thickness, so the excess thickness is not important.

In addition, by comparison of model tank performance with and without the conical roof, it was observed that the roof had little effect on tank wall deformations at mid-height sections or below. Also, the roof had no effect on the peak axial stresses measured in the tank wall (near the base) during uplifting from the sponge rubber base, while it caused a 15% reduction in these stresses during uplifting from the rigid concrete base. From these observations it has been concluded that test results obtained with the open top model can be correlated with the field behavior if the 15% peak stress reduction caused by the roof is taken into account.

COMPARISON OF ACTUAL AND SIMULATED EARTHQUAKE MOTIONS

Actual Ground Motions at the Tank Site

The M = 6.5 Coalinga earthquake had a peak recorded ground acceleration of 0.54g. This strong motion accelerogram was obtained in the switchyard at the Pleasant Valley Pumping Plant, at a distance of 5.8 miles from the epicenter. A record also was obtained on the basement floor of the pump plant, where the peak acceleration was 0.33g. Locations of these recording stations and of the epicenter are shown in Fig. 5; the three components of the recorded accelerograms and their corresponding response spectra are presented in Figs. 6a and 6b. It should be noted that the basement accelerograph was located 17 ft below grade while that in the switchyard was at the top of a slope about 70 ft above plant grade.

A series of aftershocks followed the main Coalinga earthquake, and the largest of these occurred with magnitude M = 5.1 on 9 May, by which time several additional accelerographs had been deployed. The closest of these (SMA-4) was only 0.6 miles from tank T400 whereas the record from the main shock was obtained at a distance of 7.8 miles from the tank. Figure 5 also shows the location of the aftershock epicenter and of tank T400, as well as of the 6 additional accelerographs installed to record the aftershocks. The distance of the tank from the epicenter of the main shock was about 3.9 miles while it was about 2.6 miles from the aftershock epicenter.

The location of the tank on a slope near the top of a hill is similar to the switchyard location of accelerograph SMA-1. This fact, together with similarities between the switchyard accelerogram and that recorded by SMA-4 during the aftershock suggest that the switchyard main event record is a better indication of the motions that damaged T400 than is the pump station basement record. The assumed tectonic structure of this earthquake also suggests that the switchyard record should be reasonably similar to the motions at the tank site. Eaton, et al.[9] have studied the earthquake mechanism and have concluded that it probably was a low angle thrust fracture; Bolt [10] also believes this could have been the basic mechanism. According to this hypothesis, the tank is located about 23 miles southwest of the ground surface projection of the fault, and the pumping station is at a distance of about 15 miles on the same side of the projected fault break; thus, the motions at both locations would be expected to be similar. Moreover, the low angle thrust mechanism would be expected to cause much larger horizontal accelerations than vertical, in conformance with the switchyard record.

Of course, the intensity of motion at the tank site would be expected to differ from that at the switchyard site because of their

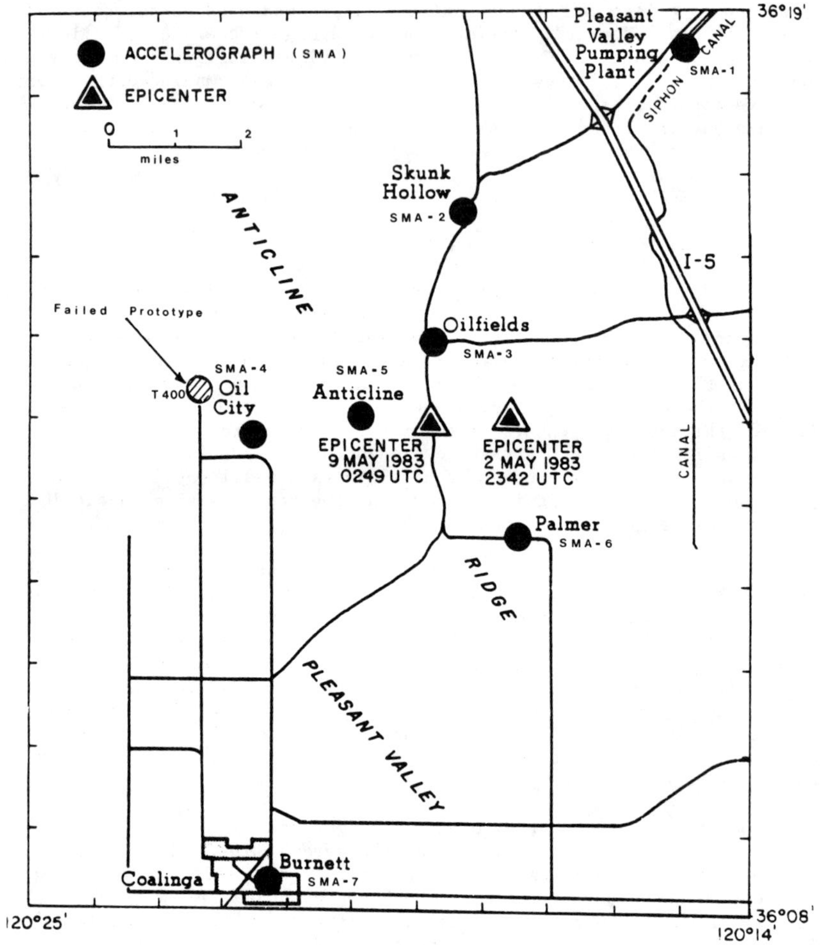

Fig. 5. Locations of epicenters and strong motion instruments for Coalinga main event and strongest aftershock together with the location of damaged tank T400. (Only SMA-1 was in location during the main event). Courtesy U.S.G.S.

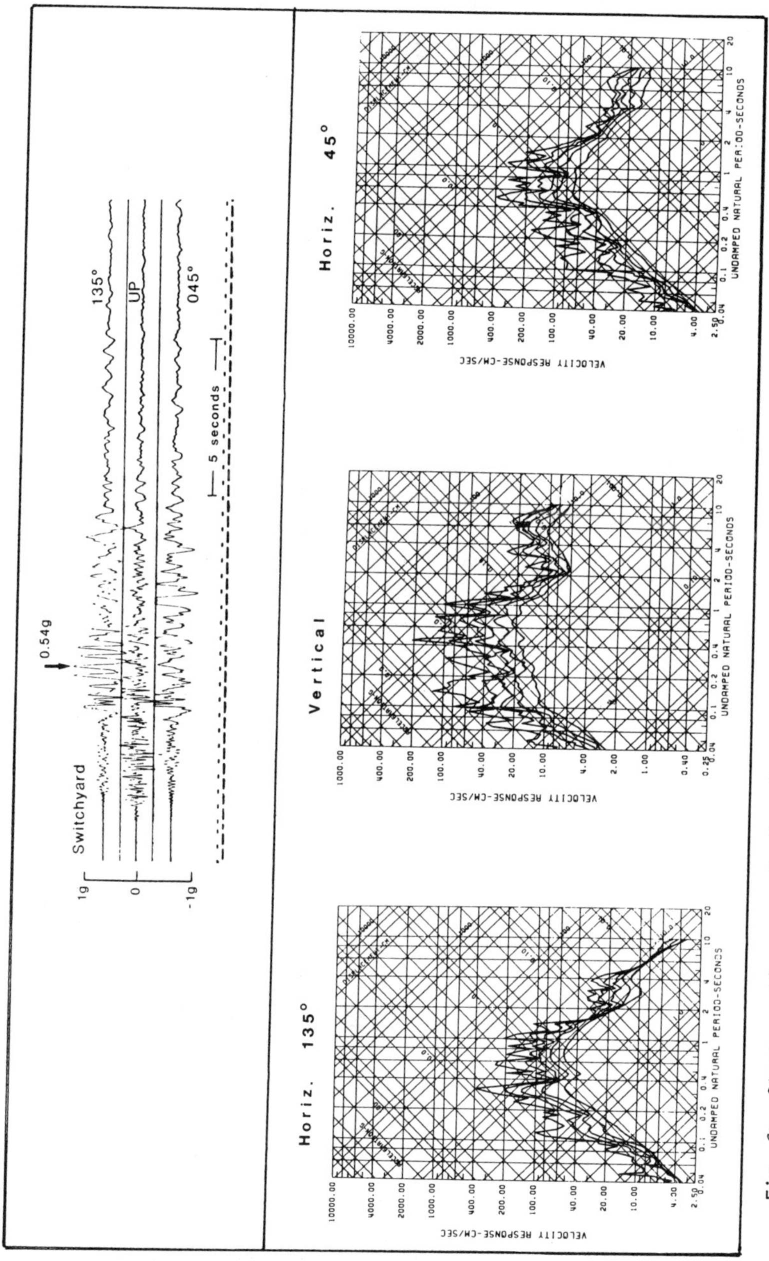

Fig. 6a. Strong motion records for the Coalinga main event at the switchyard of the Pumping Plant together with the response spectra. Courtesy U.S.G.S.

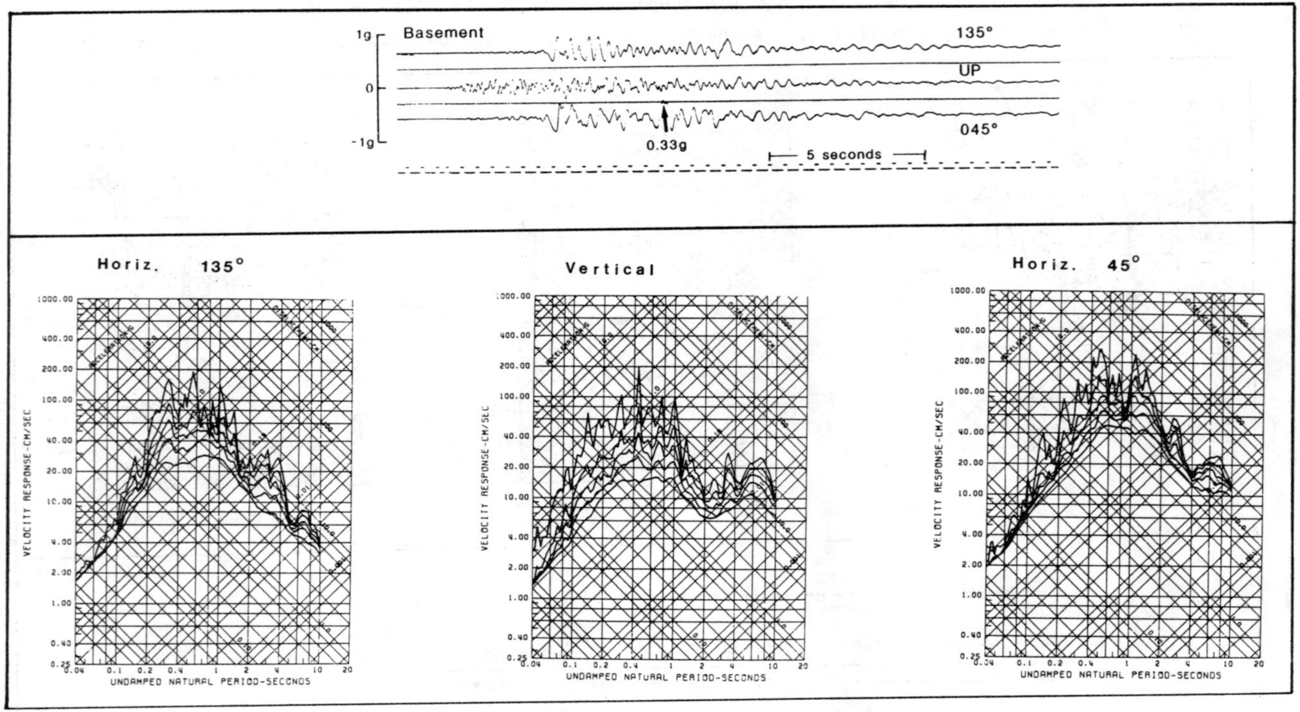

Fig. 6b. Strong motion records for the Coalinga main event at the basement of the Pumping Plant together with the response spectra. Courtesy U.S.G.S.

different distances from the earthquake epicenter. In order to evaluate the attenuation of intensity with distance, the peak horizontal acceleration data set recorded during the main (M = 5.1) aftershock by the original seismographs plus the supplementary network was studied, using data reported by Maley, et al. [11]. The eight recording stations were divided into two groups of four, according to whether or not significant local site amplification was expected, and the following attenuation formula was desired

$$a = 9.4 \; e \; 0.65 \; M \; (R - f_d + b_1)^{-b_2}$$

in which

a = peak horizontal acceleration in gals

M = earthquake magnitude = 5.1

R = hypocentral distance in miles

f_d = focal depth = 8.0 miles

Constants b_1, b_2

= 1.2, 0.5 for sites without amplification

= 0.2, 0.2 for sites with amplification.

Assuming then that the same relationship could be applied to the main ($M = 6.5$, $f_d = 6.6$ miles) earthquake, the peak acceleration was calculated for the hypocentral distances of the two pumping plant seismographs and also for the location of tank T400 (R = 7.6 miles). The calculated acceleration at the two seismograph locations were 0.54g and 0.33g, which agree well with the corresponding measured values: 0.54g and 0.35g. The calculated acceleration at the T400 tank site was 0.62g, which is 15% greater than the largest recorded acceleration during this event, and is over twice as large as was recorded by SMA-4 near the tank site during the aftershock. Based on this analysis, it was assumed that the ground motion which damaged T400 had spectral characteristics similar to those recorded at the switchyard station during the main event, but increased by 15% to give a peak N-S component of 0.62g and a peak E-W component of 0.49g.

To obtain an independent check of this result, the attenuation curves for peak rock accelerations derived by Schnabel and Seed [13] were applied to the main earthquake magnitude and the epicentral distance of the tank site. The result for the rock acceleration was 0.52g, so amplification by the surficial material to a peak value of 0.62g seems quite reasonable.

Shaking Table Test Motions

In contrast to the field situation, the motions applied to the test tank were measured precisely. Three different single component excitations were used in the tank model research, these being the earthquake simulator reproductions of the 1940 El Centro earthquake record (N-S component), the Pacoima Dam record obtained during the 1971 San Fernando earthquake (S74W component), and the 1966 Parkfield earthquake record (N65E component). All of these motions were adjusted to a peak acceleration of 0.5g as shown in Fig. 7. Note that only a single horizontal component of motion can be produced by the simulator, and that

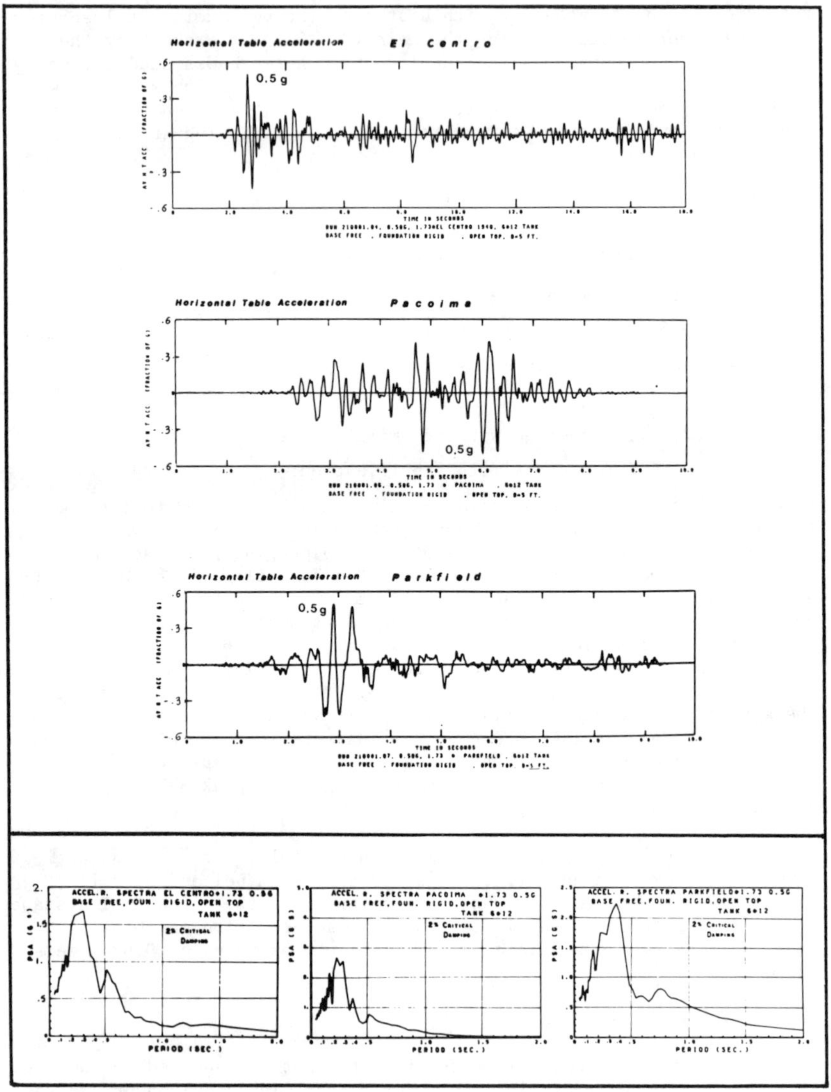

Fig. 7. Shaking table horizontal acceleration time histories recorded during the model tank tests together with the corresponding acceleration response spectra.

the vertical component was not applied in these tests because previous studies had demonstrated that it had little effect on the axial stress in the tank wall (thus, it would not affect the buckling behavior).

Acceleration response spectra for the table motions recorded during the tests also are shown in Fig. 7. The differences between these laboratory spectra and the spectra for the assumed tank site motions represented by the 135° switchyard spectrum shown in Fig. 6a were accounted for by analytical procedures to be discussed later.

DYNAMIC RESPONSE BEHAVIOR OF THE TANK

Observed Damages of Tank T400

Several photographs of Tank T400 and some adjacent tanks are presented in Fig. 8. Figs. 8a and 8b show the hillside location of T400 and of its nearly identical neighbor T600. The significant elephant's foot bulge at the base of T400 is clearly visible in Fig. 8c, and Fig. 8d shows that T600 buckled similarly but to a much lesser extent. The nearly 2 in. outward bulge at the southwest side of T400 shows prominently in Figs. 8e and 8f; the lesser degree of buckling at the northeast side of the tank is apparent in Figs. 8g and 8h. From the circumferential distribution of the buckling shown in these figures, it may be concluded that T400 was damaged by overturning moments induced by horizontal accelerations acting predominately in the northeast-southwest direction.

Performance of the Model Test Tank

The six foot high by twelve foot diameter aluminum test tank is shown during three different shaking table tests in Fig. 9. As mentioned earlier, all tests were applied with the same peak acceleration, 0.5g; the noticeably differing surface wave patterns demonstrate how the sloshing behavior is influenced by the different frequency characteristics of the three earthquakes.

These frequency characteristics also have differing effects on the stress and displacement response of the model, as is shown by the peak response values listed in Table 2. Note that results are listed for two different base conditions: free to uplift from the rigid base or from the flexible sponge rubber base. For the rigid base case, the peak stresses and displacements vary between the different excitations by factors as great as 1.85, with the Pacoima earthquake causing the greatest response. This behavior correlates well with the relative response spectra values in the period range of this tank-liquid-base system (0.24 to 0.27 sec, as indicated by response records). The high compressive axial stresses developed in this rigid base case are due to the coupled out-of-round distortions and uplift that concentrate the weight of the system in a narrow zone of the tank wall.

With flexible base, the same type of distortion produces much smaller stresses because the soft support system distributes the load over a much wider zone. In this case, the differing earthquake characteristics again cause major differences in the relative amplitudes of displacement, but the effect on the axial stresses is small. Again the

Fig. 8a. Location of Tank T400.
(mid-level)

Fig. 8b. Damaged Tank T400 together with less damaged Tank T600

Fig. 8c. Damaged Tank T400.

Fig. 8d. Damaged Tank T600

Fig. 8e. Tank T400 (SW side damage)

Fig. 8f. Tank T600 (SW side damage)

Fig. 8g. Tank T400 (NE side damage)

Fig. 8h. Tank T400 (NE side damage)

El Centro

Pacoima

Parkfield

Fig. 9. Shaking table tests of model tank

Pacoima earthquake is seen to cause the greatest response, in conformity with the response spectrum characteristics in this period range (0.23 to 0.29 seconds). It also is worth noting that the relative wave height (sloshing) response correlates well with the response spectrum values in the period range of sloshing (about 2 seconds). The Parkfield earthquake produces the largest wave heights and the sloshing amplitude is independent of whether the base is flexible or rigid.

Table 2: Maximum Response Values for Test Model

Base Condition	EQ	Displacements (inches) Top Radial	mid-Height Radial	Bottom Uplift	Compressive Stress Axial (ksi)	Hoop (ksi)	Water Pressure (psi)	Wave Height (in)
Rigid	EC	0.643	0.388	0.473	3.73	3.590	2.23	5.40
	PA	0.754	0.510	0.881	5.24	5.180	2.24	2.00
	PD	0.792	0.455	0.613	2.91	3.940	1.30	5.95
Flexible	EC	1.100	0.606	1.210	0.683	6.710	2.56	5.60
	PA	1.420	0.888	1.750	0.780	7.350	2.93	2.50
	PD	1.050	0.516	1.180	0.680	5.940	1.56	6.80

Earthquakes: EC = El Centro PA = Pacoima PD = Parkfield

The peak measured axial membrane stresses due to the three different test excitations applied with the two different base conditions are listed in Table 3. These are expressed in kips per square inch and also as a dimensionless buckling parameter $\sigma R/Et$. For comparison, the design

Table 3: Comparison of Model Tank Axial Stresses

Observed Model Test Stresses: Base Condition	EQ	Stress (ksi)	Coefficient ($\sigma R/Et$)	Specified Code Stresses: Designation	Code	Stress (ksi)	Coefficient ($\sigma R/Et$)
Rigid	EC	3.73	0.336	Max. Allowable	AWWA	1.84	0.166
	PA	5.24	0.472		API	2.23	0.201
	PD	2.91	0.262				
Flexible	EC	0.683	0.062	Critical	AWWA	2.73	0.245
	PA	0.780	0.070				
	PD	0.680	0.061		API	3.35	0.301
EC = El Centro, PA = Pacoima, PD = Parkfield				Classical	--	6.81	0.612

stresses allowed for this tank by the American Water Works Association (AWWA) and the American Petroleum Institute (API) and the nominal critical buckling stresses proposed by these organizations also are listed in the same way. The additional entry labeled "classical" is the theoretical buckling stress for a perfect cylinder subjected to uniformly distributed axial load.

This comparison shows that all three test earthquakes cause stresses in the rigid base case that are well beyond the design stress allowed by either code. Also, the stresses produced by two of the test motions exceed the nominal critical buckling stresses of both codes. The fact that no buckling was observed in any model test demonstrates the essential conservatism in the code specified stresses. The effect of the flexible foundation pad in reducing the peak axial stresses also is apparent in this table; the reductions amount to factors of four or five.

Estimation of Prototype Performance from Model Results

The axial membrane stress in the tank shell is the most significant measure of tank performance because the tendency toward buckling depends on this value. Therefore an estimate was made of the axial stress developed in the prototype from the corresponding stresses measured in the model during shaking table tests, as listed in Table 3. The extrapolation was carried out in two steps; first, the overturning moment produced in the prototype tank by the estimated Coalinga earthquake motions was compared with the scaled overturning moment in the model caused by the shaking table motions; i.e., the model moment was extrapolated to prototype scale using the standard similitude relations. Then the stress measured in the model was extrapolated to prototype scale, using the appropriate similitude relations, and this result was amplified according to the ratio of calculated prototype overturning moment to the extrapolated model test overturning moment.

Overturning Moment: The prototype overturning moment was evaluated using the standard Housner model, considering both an impulsive and a convective mass of the liquid [14]. However, following current design procedures [15,16] both masses were subjected to amplified earthquake accelerations (not just the convective mass as suggested by Housner) as indicated by the switchyard motion response spectra (Fig. 6a) increased by 15% as described above. For this purpose, the uplift deformation response period observed in the model (0.23 to 0.29 sec) was extrapolated to a prototype response period of 0.41 to 0.50 sec. The impulsive response acceleration factor (C1) was then taken as the square root of the sum of the squares of the two horizontal component 2% damped spectra. The resulting mean value acceleration in this period range was 2.82g. A similar procedure was used to obtain the effective acceleration for the convective mass (C2) using the period of the fundamental sloshing mode extrapolated to prototype scale. The effective acceleration for this 3.7 second period and 1% damping was found to be 0.082g. The combined overturning moment due to both impulsive and convective masses was 13,200 kip-ft; for comparison it may be noted that the code specified overturning moment based on a seismic coefficient of 0.24g is only 1,675 kip-ft. Moreover, the largest overturning moment to which the code procedure for stress analysis may be applied in this case is only 2,240 kip-ft. Thus, it is clear that the tank has been subjected to forces far greater than are contemplated by the standard design procedure.

Overturning moments for the test model were calculated by the same procedure, using the impulsive and convective masses for the model and the spectral coefficients derived at the model period ranges from the response spectra for the three different earthquakes (Fig. 7). These moments were then extrapolated to prototype scale using standard similitude relationships. The effective model acceleration coefficients, C1 and C2, as well as the resulting model overturning moments are listed in Table 4 for the three earthquake inputs and the two base support conditions. Also listed are corresponding values for the prototype tank based on code specifications and on the dynamic analysis procedure, as well as the ratio of prototype to model overturning moments. It is important to note that a major reason for the moments in the prototype tank being larger than the model moments extrapolated to prototype scale is the relatively greater depth of water in the prototype tank; this is in addition to the obvious effects of the more intense earthquake motions applied to the prototype as represented by the seismic coefficients, C1 and C2.

Axial Membrane Stresses: If the model satisfied the similitude conditions of Table 1 exactly, the strains in the model would be the same as those in the prototype under equivalent loading conditions, and the ratio of the prototype to model stress would be the same as their modulus of elasticity ratio. In the present case, however, corrections must be made for discrepancies in the shell thickness as well as in the elastic modulus, so the procedure for extrapolating the measured model stress to the prototype tank is as follows:

$$\sigma_p = \left(\frac{30 \times 10^3}{10 \times 10^3}\right) \left(\frac{10 \times 10^3}{9375}\right) \left(\frac{0.080}{0.056}\right) \sigma_m = 4.571\ \sigma_m$$

For the model on rigid base, the extrapolated stress must be reduced by 15% to account for the effect of the prototype roof, as noted earlier. Thus, the rigid base stress extrapolation factor is 3.885. In addition, the stress must be increased to account for the ratio of actual earthquake overturning moment on the prototype to the model moment extrapolated to prototype scale, as indicated in Table 4. Thus, the estimated stresses in the prototype tank are given by the product of the measured stress values of Table 3, the moment ratio of Table 4 and the appropriate one of the above stress extrapolation factors. The extrapolated stresses are listed in Table 5 together with the corresponding buckling stress coefficients, $\sigma R/Et$.

It is apparent in this table that significantly different stresses are indicated for the different table motions even though their individual spectral characteristics have been applied directly in the extrapolation. However, they are consistent enough that the average stress indicated for each base condition should be meaningful. Thus, the average value of 29.43 ksi is considered to represent the axial shell stress that would have been developed in Tank T400 if it had been supported on a rigid base, while 5.94 ksi represents the stress that would have been developed if the tank were on a very soft soil base. The actual prototype base condition must be between these limiting cases, and it certainly is much closer to being rigid than to a very soft soil. Thus, it is believed that 24 ksi would be a reasonable estimate of the stress that would have been developed in tank T400 if it had not buckled; of course, the actual peak stress during buckling would have been significantly less than this value.

Table 4: Tank Acceleration Coefficients and Overturning Moments

	Base Condition	EQ	Acceleration Coeffs.(g) (C1)	(C2)	Overturning Moment (kip-ft)	Moment Ratio Prot./Model
Model Test	Rigid	EC	1.665	0.070	5890	2.26
		PA	2.475	0.025	8333	1.60
		PD	1.806	0.105	6550	2.03
	Flexible	EC	1.70	0.070	6010	2.22
		PA	2.65	0.025	8920	1.49
		PD	2.07	0.105	7420	1.80
Prototype Tank	Code Val		0.24	0.082	1680	
	Max. Code		0.35	0.082	2240	
	Coalinga Eq.		2.82	0.082	13,320	

Note: Model Moments Extrapolated to Prototype Scale
Key: C1 = Impulsive Acceleration, C2 = Convective Acceleration
EC = El Centro PA = Pacoima PD = Parkfield

Table 5: Comparison of Prototype Axial Stresses

Coalinga Earthquake Stresses Extrapolated from Model				Specified Code Stresses			
Base Condition	EQ	Stress (ksi)	Coefficient (σR/Et)	Type	Code	Stress (ksi)	Coefficient (σR/Et)
Rigid	EC	32.76	1.405	Max. Allowable	AWWA	4.67	0.200
	PA	32.58	1.397				
	PD	22.95	0.984		API	4.41	0.189
	Ave	29.43	1.262				
Flexible	EC	6.93	0.297	Critical	AWWA	5.47	0.235
	PA	5.31	0.228				
	PD	5.60	0.241		API	6.61	0.284
	Ave	5.94	0.255				
Assumed Prototype Case		24.0	1.03	Classical	---	14.27	0.612

To put this predicted stress in perspective, the maximum design stresses permitted in the prototype tank by two codes, and also the nominal critical buckling stress specified for the prototype by these codes also are listed in Table 5. In addition the classical buckling stress is listed for the shell of tank T400 subjected to uniform axial compression. Comparison of these code values with the estimated Coalinga earthquake stress of 24 ksi demonstrates the severity of the ground motions developed at the tank site, and also explains why the elephant's foot buckling occurred. The inherent conservatism in the code specified stresses is apparent in this listing; but in Table 4 it also is evident that the code specified design forces are even farther on the unconservative side when compared with the effects of a significant earthquake. Apparently these two factors are intended to compensate in the code design procedure, and to lead to an effective design; but it is evident that they do not insure that the end product will be an earthquake proof tank.

SUMMARY AND CONCLUSIONS

(1) The 6 ft by 12 ft aluminum tank that has been tested in previous research programs at the University of California Earthquake Engineering Research Center is a valid structural model of Tank T400, damaged during the Coalinga earthquake; therefore stress values measured in the model can be extrapolated to the prototype scale if appropriate similitude adjustments are made with regard to discrepancies in shell thickness and modulus of elasticity.

(2) Ground motion attenuation relationships derived from records made by a strong motion accelerograph network during a strong aftershock provide a basis for estimating the acceleration intensity at the site of Tank T400 during the main Coalinga earthquake. These relationships suggest that the peak acceleration at the tank site was about 15% greater than the value recorded at the Pleasant Valley Pumping Plant switchyard site; thus, it is assumed that the switchyard response spectra increased by 15% would represent the tank site spectra.

(3) Based on these ground motion response spectra, the overturning moments calculated for Tank T400 by a modified Housner procedure are found to be nearly twice as great as the (prototype scale) moments extrapolated from model test data. A major factor contributing to the greater prototype moments is the greater depth of water in the prototype, but the more intense input motion also has an important effect.

(4) Stresses estimated in the prototype tank during the Coalinga earthquake, by extrapolation from model results, are very high when compared with design stresses allowed by typical codes; however, these design stresses are recognized to be very conservative. It must be noted in addition, though, that the estimated stresses also are very high when compared with the "classical" critical buckling stress for this tank. The buckling coefficient predicted to be developed during the earthquake is approaching twice the "classical" buckling coefficient; therefore, the observed elephant's foot buckling was to be expected.

(5) Many other tanks in the vicinity of Tank T400 undoubtedly would have developed similar buckling if they had been full at the time of the earthquake; their low liquid levels led to correspondingly low overturning moments and resulting axial stresses.

ACKNOWLEDGEMENTS

The assistance of the Shell Oil Company, Coalinga Production Division, in obtaining the field observations is hereby acknowledged. In addition, the valuable information received from the U.C. Berkeley Seismographic Station and from the U.S. Geological Survey about the strong motion seismograph records also must be recognized. Finally thanks are due to the eleven organizations under the administrative leadership of the Standard Oil Company of California for their support during the shaking table test series discussed in this paper.

REFERENCES

1. Clough, D. P., "Experimental Evaluation of Seismic Design Methods for Broad Cylindrical Tanks", University of California, Berkeley, EERC Report UCB/EERC-77/10, 1977.

2. Manos, G. C. and Clough, R. W., "Further Study of the Earthquake Response of a Broad Cylindrical Liquid Storage Tank Model", University of California, Berkeley, EERC Report UCB/EERC-82/07, February 1983.

3. Manos, G. C. et al., "Experimental Investigation of a Cylindrical Tank under Earthquake Loading", in the Proceedings of the Seventh European Conference on Earthquake Engineering, Vol. 6, pp. 233-240, Athens, 1982.

4. Manos, G. C. and Clough, R. W., "Response of a Cylindrical Liquid Storage Tank to Static and Dynamic Lateral Loads", in Earthquake Ground Motions and its Effects on Structures, AMD Vol. 53, ASME, pp. 77-90, New York, 1982.

5. Manos, G. C. and Clough, R. W., "The Measured and Predicted Shaking Table Response of a Broad Tank Model", in Earthquake Behavior and Safety of Oil and Gas Storage Facilities, Buried Pipelines and Equipment, PVP-Vol.77, ASME, pp. 14-20, New York, 1983.

6. Manos, G. C., "The Observed Earthquake Performance of Tanks During the Coalinga, California Earthquake and Correlation with some Predicted Response Values", report in preparation.

7. Shih, C. F. and Babcock, "Scale Model Buckling Tests of a Fluid Filled Tank under Harmonic Excitation", Proceedings of the Pressure Vessels and Piping Conference, ASME, San Francisco, Calif., 1980.

8. Cambra, F. J., "Earthquake Response Considerations of Broad Liquid Storage Tanks", University of California, Berkeley, EERC Report UCB/EERC-82/25, November 1982.

9. Eaton, J., "Seismic Setting, Location and Focal Mechanism", The Coalinga Earthquake Sequence Commencing May 2, 1983, U.S. Geological Survey, Menlo Park, 1983.

10. Bolt, B. A., "Seismological Observations on the Coalinga California, Earthquake of May 2, 1983", EERI Report, Berkeley, California.

11. Maley, R. et al., "Analog Strong-Motion Data and Processed Main Event Records", The Coalinga Earthquake Sequence Commencing May 2, 1983, U.S. Geological Survey, Menlo Park, 1983.

12. Borcherdt, R., "Digital Strong-Motion Data", The Coalinga Earthquake Sequence Commencing May 3, 1983, U.S. Geological Survey, Menlo Park, 1983.

13. Schnabel, P. B. and Seed, H. B., "Accelerations in Rock for Earthquakes in the Western United States", University of California, Berkeley, EERC Report No. EERC-72-2, July 1972.

14. Housner, G. W., "Dynamic Pressures on Accelerated Fluid Containers", Bulletin of the Seismological Society of America, Vol. 47, pp. 15-35, 1957.

15. "Welded Steel Tanks for Water Storage", AWWA Standard D100-79, American Water Works Association, Denver, Colorado, 1979.

16. "Welded Steel Tanks for Oil Storage", API Standard 650, Sixth Ed., Revision 3, American Petroleum Institute, Washington, D.C., October, 1979.

INTERMOUNTAIN POWER PROJECT SEISMIC DESIGN

By Luis E. Escalante,[1] Robert P. McBean,[2]
P. Bradford Vaughan,[3] and Arthur W. Shively,[4]
Members, ASCE

ABSTRACT: Coal-fueled generation units of the Intermountain Power Project are being constructed in a seismically active area of Utah (UBC Zone 3). The seismic design criteria were published during the initial design stage (See ASCE Journal of Technical Topics, April, 1983). This paper describes the implementation of those criteria to detailed design, interaction with equipment suppliers, and design insights.

INTRODUCTION

The two 761 MWe coal-fueled electric generating units of the Intermountain Power Project (IPP) will supply power to Utah and eastern Nevada, and through a ±500 kV dc Southern Transmission System line, to the Los Angeles area. The station is being constructed near Delta, Utah, in an active seismic area (UBC Zone 3), and is believed to be the first major coal-fueled plant in the United States to be fitted with broad-based seismic protection. Commercial operation of Units 1 and 2 is scheduled for 1986 and 1987, respectively. Design engineering began in 1980 and construction began in 1981. The completed station is pictured in Fig. 1.

The philosophy, criteria, and procedures used for seismic design of the structures and equipment were presented in an earlier paper (1). Described herein are the implementation

FIG. 1. – Intermountain Power Plant

[1]Manager, Civil Structural Engineering Design and Construction Div., Dept. of Water & Power, City of Los Angeles, Calif.

[2]Supervisor, Structural Analysis Group, Power Div., Black & Veatch, Engineers-Architects, Kansas City, Mo.

[3]Structural Engineer, Power Div., Black & Veatch, Engineers-Architects, Kansas City, Mo.

[4]Structural Engineer, Power Design and Construction Div., Dept of Water & Power, City of Los Angeles, Calif.

of those criteria in detailed design engineering, the interaction of the project manager (Los Angeles Department of Water & Power) for the participating utilities and the consulting engineers (Black & Veatch) with equipment suppliers to achieve compliance with the specifications, and certain design insights acquired.

SEISMIC DESIGN CRITERIA

General. – Studies of the geology and seismology at the site indicated that the plant may be exposed to an earthquake with peak ground acceleration of 0.2g (1). Seismic design criteria were developed to limit service interruptions from this important power source by minimizing the seismic damage to those structures and equipment essential to power generation. In general, more stringent criteria were selected when they offered a substantial increase in reliability for a minimal cost increment.

Buildings. – Plant buildings were seismically designed in accordance with the Uniform Building Code (UBC), 1979 edition (6). The base shear, V, was computed from V = ZIKCSW, but not less than 0.15W for buildings essential to power generation. Zone coefficient, Z, was taken as 0.75; importance factor, I, was taken as 1.0 to 1.5 (the latter for buildings essential to power generation); force factor, K, was taken as 1.0 for braced steel buildings; the fundamental period factor, C, was taken to be less than or equal to 0.12 (based on an estimate of the period for each building); soil-structure interaction factor, S, was taken as 1.2 for the dense sands and very stiff clays at the site; and the weight, W, included the total dead load of the structure and the weight of equipment and loads which are likely to be present during an earthquake ("normal" operating levels of water, coal, ash, etc.).

Equipment. – Most equipment was procured using detailed specifications prepared by Black & Veatch for installation by others in various construction contracts. Certain major components were purchased under furnish/erect contracts (e.g., steam-generator, air quality control system, cooling towers). Specifications therefore included appropriate criteria for the manufacture and anchorage of all equipment. Detailed specifications required application of engineering judgment with regard to the following: seismic withstand capability of standard designs, especially with regard to anchorages and appendages; modifications necessary to improve the seismic resistance of standard designs; selection of ductile materials; safety concerns for operating personnel; significance of the component relative to continued operation of the plant; potential cost of repair or replacement and associated outage duration after an earthquake; availability of replacement parts or alternate systems; resultant effects of components on each other during an earthquake; initial construction cost; and expected supplier compliance.

DESIGN OF THE GENERATION BUILDING

Building Description. – The generation building consists of the turbine, boiler or steam-generator, and fan areas. On the south side of the generation building is the turbine area, measuring 350 feet (107 m) long by 104 feet (32 m) wide by 93 feet (28 m) high. This area encloses the turbine-generator and related equipment. The turbine-generator is supported on a seismically isolated reinforced concrete pedestal, integrally constructed with pedestals for the two boiler feed pumps. The steam-generator area, measuring 230 feet (70 m) long by 216 feet (66 m) wide by 303 feet (92 m) high, houses the boiler itself, the steam drum, coal silos, conveyors, the deaerator, heaters, and associated equipment and piping, all of which are supported above grade, as well as coal pulverizers and other equipment on the ground floor. The fan area, located north of the steam-generator area and measuring 230 feet (70 m) by 93 feet (28 m) in plan, with various roof heights, houses the air preheaters, forced draft fans, and related ductwork. The three areas of the generation building comprise a single structure and are separated functionally, although not physically. The turbine area gains most of its lateral resistance from the heavily braced steam-generator area. To minimize the seismic interaction of the two generation buildings, the turbine areas of Units 1 and 2 are seismically isolated from each other. A double set of columns is used at the interface, and roof bracing and floor systems are not contiguous.

Structural engineers for the steam-generator supplier, Babcock & Wilcox, began working with their counterparts at Black & Veatch immediately after award of the contract in order to achieve a good seismic framing system within the furnace and at interfaces with the building; this cooperation early in the design was essential.

Seismic Design. – The generation building was analyzed and designed as a braced frame in accordance with the seismic provisions of the UBC. The total base shear was computed to be approximately 25,000 kips (111,250 kN). In contrast, the base shear resulting from an 84 mph (38 m/s) design wind was computed to be about 5,200 kips (23,100 kN).

The weight of the steam-generator is second only to the weight of the building itself. Plate girders 20 feet (6.1 m) deep suspend the boiler from 290 feet (88 m) above grade. Horizontal seismic forces are tranferred at six levels from buckstays to the building steel by seismic ties. These ties consist of pinned links which transfer axial force yet can accommodate the thermal growth of the boiler. Where possible, these ties were located to coincide with major floors to transfer the forces directly into a horizontal truss. The steam drum is restrained from side-to-side motion by centrally located bumpers, and from fore-and-aft motion by snubbers at each end of the drum.

The base shear was distributed to the floors in accordance with UBC requirements. (The boiler weight was apportioned to those floors to which it is seismically tied.) At each floor, the horizontal seismic force was distributed over the floor in proportion to the weight distribution. The floor beams and horizontal bracing were designed to transmit to the vertical bracing system the greater of the seismic forces computed from this distribution, or the seismic forces for the boiler or equipment. The intent of this provision was to assure adequate local strength in the vicinity of the boiler ties and equipment supports. The vertical bracing system, however, was designed to carry only the distributed base shear forces.

The principle braced column rows, identified in Figs. 2 and 3, are Rows 101, 103, 107, and 109 in the north-south direction, and Rows F, G, H, L, N, and R in the east-west direction. The seismic forces from the boiler are equal to its tributary operating weights times 0.20, the specified horizontal seismic coefficient. For example, each of the top two levels of ties carry north-south seismic forces of 1,000 kips (4,450 kN). These forces, together with those from the lower four tie levels, are resisted primarily by Rows 103 and 107, resulting in large vertical bracing and column uplift forces.

For the seismic analysis the building was modeled as a series of plane truss subassemblies. A stiffness analysis of each braced column row provided the values for the "springs" (or elastic restraints) that were later used in the horizontal truss analyses to represent the supports provided by the vertical bracing system. By analysis of the floors from the top of the building down, the reactions from each floor were input as loads to the next lower floor level. This procedure accounted for the collector forces present in the beams of each column row without the need for additional computer runs. Following analysis of the final (lowest) floor, each level of each vertically braced column row was displaced an amount equal to the corresponding deflection resulting from the floor analyses, thereby providing the vertical bracing system forces.

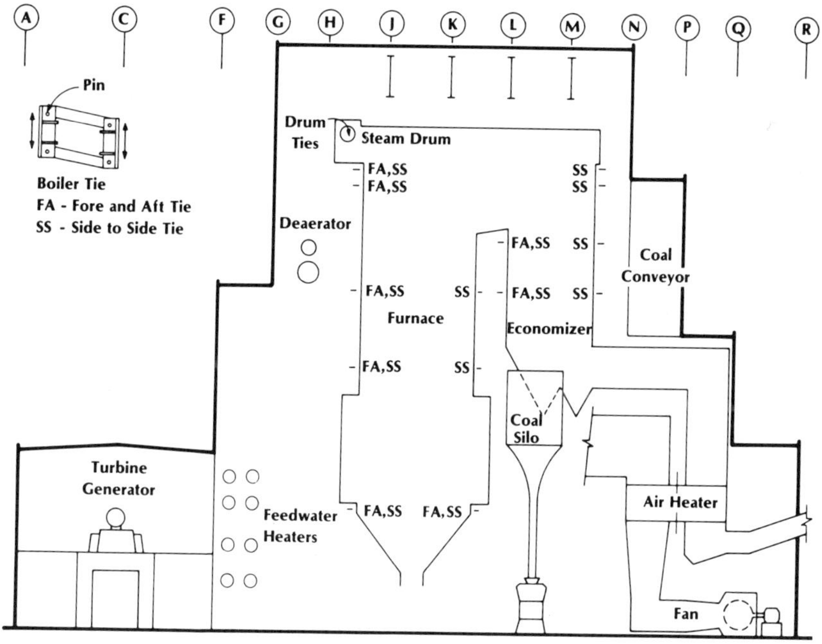

FIG. 2. – Elevation View of Generation Building

Following the lateral load analysis, floor members were analyzed and designed for gravity dead and live loads plus seismic forces using a floor grid design program. This program automatically downloaded the axial forces from the horizontal truss analyses and combined them with user-input gravity loads. For each member a shape was selected by the program. Relevant information regarding loading, stress levels, deflection, and controlling load cases was output. Then the gravity loads were distributed to appropriate columns for member selection by a column analysis and design program.

All member connections were designed by Black & Veatch. For resisting the large axial forces in the beams of column rows, together with vertical forces, a hierachy of member end connections was prepared. Clip angle connections were used throughout until their capacities were exceeded. When the capacity of 1 inch (25 mm) thick angles was inadequate, an endplate was used. To provide for still higher loads, an extended endplate was developed wherein the depth of the member was increased by welding a tee to the top and/or bottom flange(s) and attaching a full depth endplate. Both single and double rows of bolts were used on endplates and extended endplates. When the load-carrying capacities of the extended endplate connections were exhausted, fully welded connections were used.

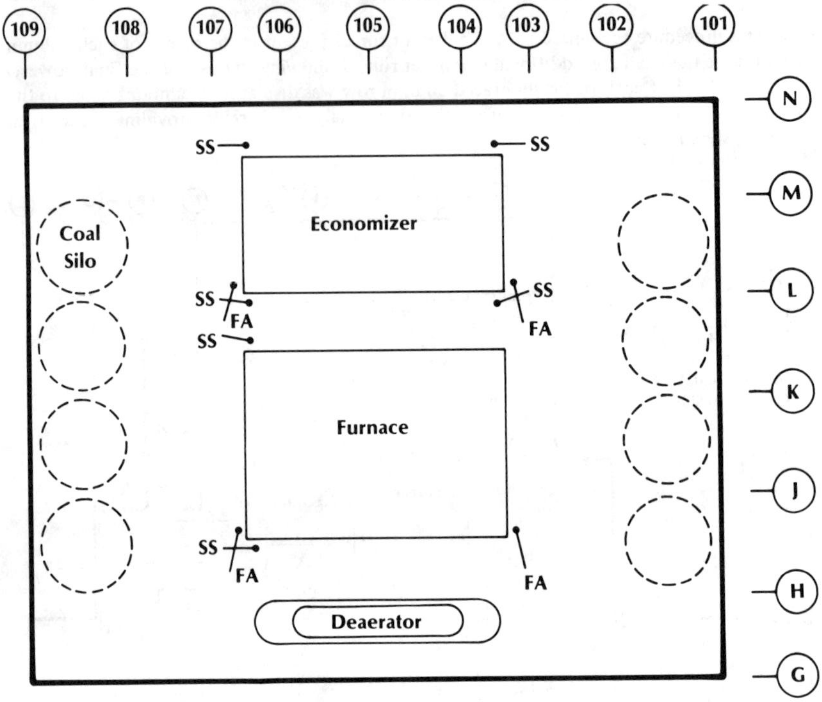

FIG. 3. – Plan View of Generation Building at Deaerator Platform

To reduce secondary bending effects, eccentricities were minimized or eliminated at all connections. This design decision contributed to the large dimensions of many vertical gusset plate connections. For vertical bracing connections the workpoint was set at the intersection of the centerlines of the column and the beam. Especially long horizontal legs on the gusset plates resulted. It may be more economical to locate the work points at the top/bottom of beam – centerline of column intersection and design for the resulting eccentricities. Figures 4a and 4b illustrate the effect of relocating the workpoint on the dimensions of the gusset plate.

The UBC requirement that connections of members in seismic lateral load-carrying trusses must be designed for dead load plus live load plus 1.25 times the seismic forces without a one-third increase in the allowable stresses also contributed significantly to the size of connections. If for a particular connection the calculated wind and seismic forces were identical, the required number of bolts for the seismic case would be approximately 1.25 x 1.33, or 1.67, times as great as the number of bolts for the wind case. Similarly, the dimensions of the gusset plates for the seismic case would be much larger than for the wind case. Figures 4b through 4d illustrate the effect of the UBC requirements for large vertical bracing gusset connections. In each case, the basic force in the brace is assumed to be 1,178 kips (5,238 kN). If the basic force is multiplied by 1.25 as required for seismic load, and no stress increase is permitted, the design of Fig. 4b results. Figure 4c shows a hypothetical seismic case where a 1/3 increase in the allowable stresses is applied after multiplying by 1.25. If a 1/3 increase in the allowable stresses is permitted as for wind load, the basic force is multiplied by 0.75 and the design of Figure 4d results. Figures 4b and 4d thus demonstrate the contrast of the UBC provisions for connections for wind and seismic load.

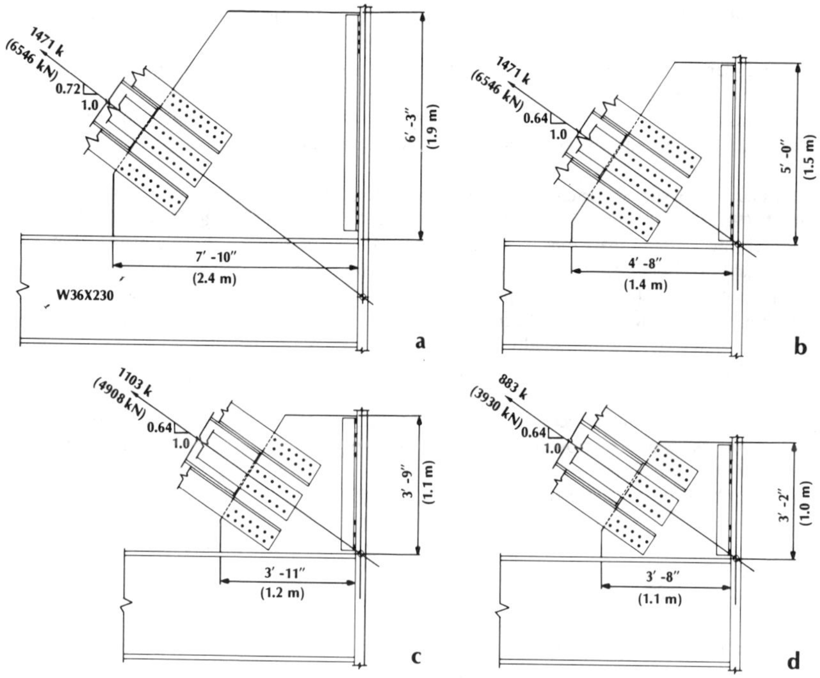

FIG. 4. – Bracing Connection Designs

DESIGN OF OTHER STRUCTURES

Other Buildings. – All other buildings for IPP, such as the control, water treatment, fabric filter, wet scrubber, coal crusher, limestone preparation, administration, and general services buildings were seismically designed to meet the provisions of the UBC. For the latter two buildings the importance factor, I, was taken equal to 1.0, whereas for the others it was taken as 1.5. No particular design problems were encountered.

Elevated Structural Components. – Conveyor galleries, coal silos, ductwork, hoppers, tanks, and similar structural components which are generally supported by a structural framework were seismically designed as follows. The horizontal seismic forces were computed by multiplying the weights by ±0.30, and the concurrent vertical seismic forces were computed by multiplying the weights by ±0.20. The weights included normal operating levels of contents. These seismic forces were distributed throughout the component in proportion to the weight distribution. The horizontal seismic force is similar to that calculated from the UBC for parts of structures (with I equal to 1.5, and C_p equal to 0.3). A vertical seismic force, not specified in UBC, was also imposed to account in a simplified way for amplification of the ground motion by the support structure and in the case of the conveyor galleries for the vertical response of the gallery itself. Anchor bolts and associated anchorage material were conservatively designed for horizontal and vertical coefficients of 0.5 and 0.35, respectively.

The wind and horizontal seismic forces on the single conveyor galleries were about the same. Except for some longitudinal seismic effects there was little design impact. For dual conveyor galleries the seismic lateral force exceeded that for wind, but generally the combination of dead plus live plus snow plus coal loads exceeded the corresponding forces from dead plus coal plus vertical seismic plus horizontal seismic loads. Therefore, only the bottom horizontal bracing and the anchorage pins were governed by seismic combinations.

The eight coal silos in each steam generation building represent a significant mass. Each 30 ft. (9.1 m) diameter silo stores 33,200 cu. ft. (930 m^3) of crushed coal. The effects of the seismic design provisions were negligible on the design of the platework, but were significant in design of the lateral support of the silos. In addition to the primary support at the bottom of the cylindrical portion of the silo, two levels of seismic ties were used to restrain lateral movements and reduce the overturning moments. Each level consisted of four pipe ties extending radially from stiffening rings on the shell to adjacent building columns at major braced floor elevations.

The seismic provisions affected the sizing and support of air ductwork very little. Flue gas ductwork, especially sections prone to collection of dust, required lateral and longitudinal restraints and strengthening over a nonseismic design. Seismic ties were located where they did not inhibit thermal movement of the ductwork.

Conveyor Bents, Tank Supports. – Two and four legged braced conveyor and tank support bents were designed for horizontal and vertical seismic coefficients of 0.35 and 0.25, respectively. Seismic forces usually governed over wind forces in the design of the conveyor bents, and for the taller towers there was considerable design impact. However, consideration of the importance of these structural components to continued operation of the plant led to the use of these coefficients rather than the less conservative UBC factors for elevated tanks.

Tanks at Grade. – Cylindrical, flat-bottom tanks for storage of oil and certain other liquids where hydrodynamic effects were judged to be significant were seismically designed in accordance with API Standard 650 (7), Appendix E, using zone coefficient, Z, equal to 0.75, essential facilities factor, I, equal to 1.5, and soil profile coefficient, S, equal to 1.2. Tanks for water storage were designed in accordance with AWWA D100 (4), Appendix A, using Z and S as previously indicated, and a structure coefficient, K, equal to 2.00. All tanks were anchored to their foundation, and suitable flexibility was provided all piping attached to the tank shell.

Foundations for Turbine-Generators and Coal and Limestone Crushers. – In addition to designing for normal operating loads and unbalance, and in the case of the turbine foundation for accident conditions, the foundations were designed for seismic ground motion. These structures were modeled as space frames with masses lumped at the joints, and seismically analyzed by the response spectrum method, using as input the 5% damping spectrum given in Fig. 5. Inasmuch as a vibration analysis is normally performed to determine the potential for undesirable vibration levels due to machine unbalance, the additional effort to perform a seismic analysis was not significant. Additional reinforcement was required to accommodate the seismic forces in the columns and some beams.

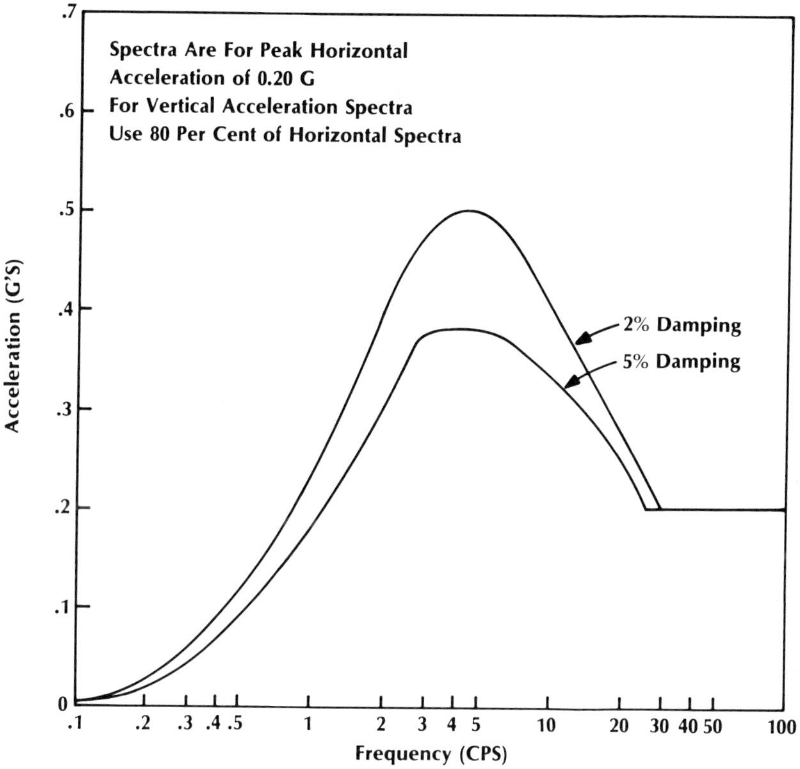

FIG. 5. – Design Response Spectra

Chimney. – The chimney is constructed of a reinforced concrete shell which supports two fiberglass reinforced plastic liners 28 ft. (8.54 m) in diameter. The liners are supported by the shell: vertically and laterally at three levels by steel framing, and laterally by stay rods at other levels. For the seismic design, the chimney contractor, Crown Union, performed a response spectrum analysis for horizontal ground motion using the 2% damping spectrum of Fig. 5 and a beam element "stick" model which incorporated the shell, liners, the lateral tie system, and a rotational spring which accounted approximately for the soil-structure interaction. Furthermore, the seismic shears and moments computed from ACI 307-79 (3), with zone coefficient, Z, equal to 1.0, and use factor, U, equal to 2.0, were used as minimum values in the shell design. For the lower 400 ft. (120 m) the ACI 307 moments governed. Above that level the vertical reinforcing design was governed by the moments from the dynamic analysis, although the difference between the moment results by the two methods was not great.

Cooling Towers. – The mechanical draft cooling towers were structurally designed by the contractor, Ceramic Cooling Tower Company, to meet provisions of the UBC, for which K was taken as 1.33. Furthermore, the contractor estimated the deflections of the tower from a simplified response spectrum analysis using the 5% damped spectrum of Fig. 5. For these seismic deflections the contractor was required to ensure the structural integrity of the tower. For each generation unit two cooling towers are required, each about 220 ft. (67 m) in diameter and over 50 ft. (15 m) tall. Many of the components are of precast concrete, so particular attention was paid to connection details. Structural engineering services were provided for the contractor by the Austin Company. The latter conducted 1/2 scale static tests of several connection types in order to evaluate construction procedures as well as connection integrity. In addition, Southwestern Research Institute conducted shake table testing of the spaced tile fill configuration. These tests, together with detailed analyses, demonstrated that the tower, comprised largely of precast units but with some cast-in-place shear walls, should perform satisfactorily in a major earthquake.

Architectural Components. – The seismic design requirements for architectural components reflected those of both UBC and ATC 3-06, corresponding to the former where applicable, and in the format of the latter for miscellaneous components. Precast concrete nonbearing and nonshear wall panels, and their connections, were seismically designed in accordance with Section 2312(j)3C of the UBC.

DESIGN OF EQUIPMENT FOR THE POWER PLANT

Procurement Specifications. – A standard seismic design specification section was developed for inclusion in all procurement specifications and in those construction specifications where the supplier performed aspects of the design. This standard section had subsections on equipment, piping, structures, architectural components, piping, and cable tray systems. Only those subsections applicable to the subject procurement were included in the specification. Once prepared, this standard section was not modified for each procurement. The standard seismic specification was of necessity rather general in nature, providing guidelines and seismic coefficients, but not referring to a specific component. Other technical sections of the specification, which were unique to the component to be purchased, pointed to the standard specification and provided the necessary seismic classification and additional seismic design requirements or clarifications. This procedure worked very well for the project, and avoided the writing and reviewing which would be required if all seismic requirements had to be tailored for each specification.

In many cases the level of effort, expected seismic performance, cost impact of what might be required by the specifications, and particular concerns were discussed with the suppliers during preparation of the specifications. Many suppliers employed consultants to assist in the seismic design work, particularly when they had no structural or mechanical engineers on their in-house staff. In general, the response by the suppliers has been very good.

Equipment Design Requirements. – In recognition of the relative seismic behavior and importance of different types of equipment, a seismic classification and performance level system patterned after the ATC 3-06 recommendations (5) was developed. Performance levels S, G, L, and N, were assigned to define the design objectives as a measure of resistance to seismic forces and displacements, as indicated in abbreviated form in Table 1. These performance levels were purposely selected to be similar to those of ATC 3-06. The technical sections of the procurement specification defined the performance level to be used by the supplier. When the procurement involved numerous components, the performance levels were called out for each component except when it was appropriate to use the same level for all components.

TABLE 1. – Seismic Performance Criteria

Performance Level	Performance Factor	Design Objective
Superior (S)	1.5	Superior resistance to seismic forces and displacement. Retain complete structural integrity. No broken piping. Should remain operable during and following an earthquake.
Good (G)	1.0	Good resistance to seismic forces and displacements. Retain complete structural integrity. No broken main piping or vessels. No shorted or broken electrical circuits. Return to service following visual inspection and, if necessary, minor repairs.
Low (L)	0.5	Low to moderate resistance to seismic forces and displacements. Sustain moderate damage but remain in place. Broken distribution piping and vessels tolerated. Return to service following repairs which may require up to 30 days.
None (N)	0.0	Low resistance to seismic forces and displacements. Anchored to remain in place.

The seismic force on each component was statically computed by concentrating the weight of the component at its center of gravity and multiplying the weight by the performance factor times ±0.30 for any horizontal force, and by the performance factor times ±0.20 for the vertical seismic force. The distribution of the seismic force was in accordance with the distribution of weight throughout the component. The resulting seismic internal stresses were combined with the stresses due to normal operating loads. For design of the anchorages, the horizontal and vertical seismic forces were taken equal to the performance factor times ±0.50 times the weight of the equipment, with no increase in allowable stresses. For most equipment the cost associated with enhancing the anchorage was minor. Conduit and piping connected to equipment were designed to accommodate anticipated relative displacements between attachment points. For most equipment the seismic criteria posed no design complications. Typical performance levels and the associated relative cost impact of the seismic provisions are provided in Table 2.

TABLE 2. – Performance Levels for Selected Equipment

Equipment	Performance Level	Relative Cost Impact of Seismic Provisions
Fans (hardmounted), Motors, Pumps	G	Minor
Fans on vibration isolation mounts	G	Minor, retaining brackets or snubbers added
Battery racks for UPS	S	Measurable, bracing and retaining bars added
Electrical cabinets	G	Minor, anchor bolt system required

Heaters	G	Minor, retaining bars needed at expansion supports
Water treatment tanks on support legs	G	Measurable, bracing needed, lateral support needed at an upper level on tall tanks
Emergency generator	S	Little to none, appendages well secured
Coal pulverizers	G	Minor
Turbine lube oil system	S	Measurable, restraints, extra pipe restraints and flexible couplings needed

DESIGN OF EQUIPMENT FOR CONVERTER STATIONS

General. – The Southern Transmission System, capable of transmitting 1,600 MWe of power, consists of two alternating current (AC) switchyards and two direct current (DC) converter stations which are connected by a ±500 kV DC transmission line 500 miles (800 km) long. Each station is capable of converting power from AC to DC or from DC to AC. For transmission from the power plant to the Adelanto site near Los Angeles, the power is converted from 345 kV AC to ±500 kV DC. At Adelanto, pictured in Fig. 6, the power is inverted from ±500 kV DC to 500 kV AC.

The two converter stations represent the state-of-the-art in high voltage DC (HVDC) transmission. The supplier, ASEA, was awarded a turnkey contract for the design and construction of both stations. The Los Angeles Department of Water & Power, acting as engineering manager for the project, defined strict design criteria and performance requirements and reviews all aspects of design and construction. The work of ASEA includes furnishing and installing two complete converter stations, DC switchyards, and associated equipment support structures and foundations. From the seismic design standpoint the most important features are the valve halls in which three 56 ton (50,800 kg) valves are suspended from the roof girders of the building, and the electrical yard equipment.

FIG. 6. – Adelanto Converter Station

Valve Halls. – The conversion of AC to DC for transmission is accomplished by stacking solid state thyristors in modular units to develop the desired voltages and currents. The assembly of these thyristor modules, called a valve, is approximately 50 ft. (15 m) tall. The valves are suspended from a roof system of welded steel girders and horizontal trusses. The hangers form a damped pendulum system, as illustrated in Fig. 7. ASEA proposed this flexible mounting approach to isolate the valves seismically and thereby minimize horizontal accelerations. The problem with base-mounted designs is that to minimize seismic motions of these tall valves the supporting structures have to be so rigid as to be economically and perhaps physically impracticable.

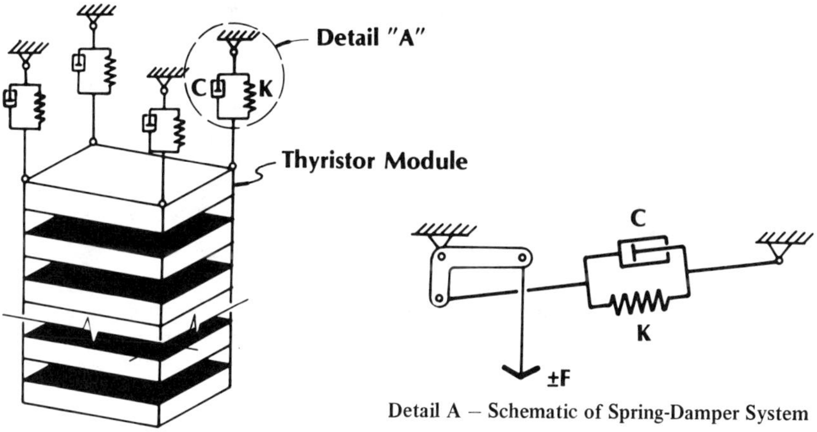

Detail A – Schematic of Spring-Damper System

FIG. 7. – Mathematical Model of Pendulum Suspended Valve

URS/Blume provided review and verification of ASEA's conceptual and analytical approach to the suspended valve design and CH_2M Hill designed the braced steel valve buildings. Previous scale model shake-table testing of the valve, support assembly, and the building had been provided by ANCO Engineers for the Sylmar Pacific Intertie Project, a project which utilized a similar but smaller version of the suspended valve concept to be installed for IPP. That testing verified ASEA's analytical approach and confirmed that no nonlinear behavior need be considered in the valve analyses.

The dimensions of the valve hall buildings were dictated by electrical clearances and the anticipated seismic displacements of the valves. Using seismically induced accelerations and displacements estimated by Blume, CH_2M Hill developed a preliminary building design. Blume then modeled and analyzed the building, investigated the interaction of the building and equipment, and provided member stresses and displacements to CH_2M Hill, who used this information to develop the final design of the building.

Converter Station DC Electrical Yard Equipment. – The DC yard equipment is typical of most high voltage switchyards. The station includes circuit breakers, current transformers, capacitor voltage transformers, surge arrestors, and bus supports. All equipment for the converter stations had to meet stringent seismic design requirements. For the Intermountain station the response spectra in Fig. 5, which have a zero period acceleration of 0.2g, provided the seismic design basis. For the Adelanto station the spectral accelerations were normalized to 0.5g. Damping was to be taken as 2% of critical damping unless a higher value could be substantiated by tests. In most cases, 5% damping was justified by the suppliers.

The seismic design of the equipment was critical because of the long, slender insulators which are so vulnerable to earthquake damage. Most vendors were able to meet the Intermountain criteria with their standard equipment, but to meet the Adelanto criteria many had to modify their designs. These modifications utilized high strength insulators, increased damping systems, and/or softening of the supporting structures.

The supporting structures were softened to lower the natural frequencies which are most subject to amplification by using Belleville washers or other compressible materials in the design of critical connections. Although these softening methods met the specification requirements, there are serious concerns about the validity of this approach. At least one earthquake, the Miyagi-ken-oki earthquake which occurred on June 12, 1978, near Sendai, Japan, has caused damage to porcelain components of equipment which had relatively low fundamental frequencies. The low frequency energy content of earthquake motions and the low frequency response characteristics of electrical equipment need to be addressed by further testing and research. However, the tests indicated that softening of a system usually results in significant increases in damping. The seismic withstand capability of the equipment should thereby be greatly enhanced.

AC Switchyard Electrical Equipment. – The AC switchyard equipment is similar to the DC yard equipment and had to meet the same stringent seismic design standards. In addition, the AC equipment was also required to pass physical vibratory tests, thereby eliminating disagreements concerning the validity of modeling techniques and assumptions used in dynamic analyses. The Adelanto converter station is located near the San Andreas fault. Since it is economically imperative that the station have a high level of reliability, the expense incurred by testing the equipment was considered prudent. The input beat motion for these tests consisted of a sinusoid of the equipment resonant frequency modulated by a lower frequency sinusoid which provided at least 10 cycles of resonant frequency per beat. The magnitude of the input acceleration was specified to be 0.2g for the Intermountain switchyard and 0.5g for the Adelanto switchyard for all natural frequencies.

Most vendors were able to meet the Intermountain criteria. However, the Adelanto criteria caused major problems for several vendors. Live tank circuit breakers could not qualify for the Adelanto site and several design iterations were required for the disconnect switch supplier to meet his contractual obligations.

Testing of the equipment revealed certain characteristics which appeared to be consistent for these flexible assemblies: the natural frequency of the system as determined by testing tended to be about 25% lower than that found by analysis; combining vertical and horizontal input motions yielded a lower response than found from horizontal input motion alone; high amplitude motions caused oval-shaped oscillations which effectively dissipated energy; and high amplitude displacements caused a frequency shift, thereby detuning the system. Consequently, the magnification factor for sine beat quasi-resonance proved to be about 3.5 to 4.5 rather than 7.5 as predicted for 5% of critical damping.

Numerous protests were registered by vendors because the prescribed sine beat test was more severe than the random motion test of IEEE 344 (2). However, the test requirements were no more stringent than those specified for electrical equipment used in the high seismicity areas of Japan. The Japanese standards require that all equipment be subjected to a 0.3g, 3 cycle sine wave input motion while limiting the resultant stress in the porcelain to 50% of the ultimate breaking strength of the porcelain as defined by the average breaking strength less three standard deviations. Their tests do not require that the equipment be tested to near failure, an advantage because of the danger and expense of breaking these large pieces of equipment.

DESIGN OF INTERMOUNTAIN-ADELANTO TRANSMISSION LINE

The ±500 kV DC transmission line will stretch for approximately 500 miles (800 km) from the Intermountain plant in Utah to Adelanto, California. About 2,000 self-supporting, lattice type, galvanized steel towers will be needed. These towers have been designed to withstand 100 mph (45 m/s) winds and large longitudinal construction loads. Experience indicates that seismic design is unwarranted for free-standing lattice towers which have been designed for moderate wind and unbalanced longitudinal loads. However, transmission towers will not be located directly on a known fault or fault splay because any distortion of the base of the tower could result in major damage or collapse of the tower.

SUMMARY

To achieve a high degree of reliability, the Intermountain Power Project has been designed to incur minimal damage and service interruptions as a consequence of seismic events. In general, the buildings have been designed in accordance with the Uniform Building Code. The UBC design requirements for the generation building included a 1.25 seismic force multiplier for members which are part of the lateral load resisting system and the requirement for no increase in allowable stresses for connections. These requirements, coupled with an effort to minimize connection eccentricities, resulted in heavy bracing and large gusset plates. These large gusset plates were expensive, difficult to erect, and at times inhibited routing of piping, cable trays, ductwork, and access walkways. If all interference consequences of the seismic provisions had been anticipated, and the design schedule had permitted, the effects of eccentric connections, redistribution of seismic forces between bays, and alternate bracing configurations could have been studied. However, the design and construction have proceeded relatively smoothly although schedule constraints did not permit extensive design time.

Equipment for the plant has been designed in accordance with detailed specifications which reflect past behavior of equipment during earthquakes, the provisions of the Uniform Building Code, the ATC 3-06 recommendations, and communication among all parties involved regarding the expected performance and cost of the seismic provisions. The standard seismic design specification concept for equipment worked well, and in most cases the requirements had minor impact on the cost of the equipment.

Detailed analyses and scale model dynamic testing of the converter station valves led to an innovative damped pendulum support system to minimize horizontal accelerations. Electrical yard equipment often required high strength insulators and softening of the supporting structures. Certain seismically vulnerable AC switchyard equipment was subjected to vibratory testing in addition to detailed analyses to assure a high level of reliability. The testing revealed that the natural frequency of much of this equipment was about 25% lower than predicted by analysis, and under large displacements a frequency shift occurred, thereby detuning the system. No seismic design was required for the lattice type transmission towers.

APPENDIX – REFERENCES

1. McBean, R. P., Escalante, L. E., and Shively, A. W., "Intermountain Power Project Seismic Criteria," *Journal of Technical Topics in Civil Engineering*, ASCE, Vol. 109, No. 1, April, 1983, pp. 73-87.

2. *Recommended Practices for Seismic Qualification of Class 1E Equipment for Nuclear Power Generating Stations*, Institute of Electrical and Electronics Engineers, 344, 1974.

3. *Specification for the Design and Construction of Reinforced Concrete Chimneys*, ACI 307-79, American Concrete Institute, 1979.

4. *Standard for Welded Steel Tanks for Water Storage*, ANSI/AWWA D100, American Water Works Association, 1979.

5. "Tentative Provisions for the Development of Seismic Regulations for Buildings," Applied Technology Council, Special Publication 510, National Bureau of Standards, also Publication ATC 3-06, Applied Technology Council, June, 1978.

6. *Uniform Building Code*, International Conference of Building Officials, Whittier, Calif., 1979.

7. *Welded Steel Tanks for Oil Storage*, ANSI/API Standard 650, American Petroleum Institute, 1980.

AN OVERVIEW OF EARTHQUAKE DESIGN REQUIREMENTS AT PACIFIC BELL

By Keith E. Johnsen P.E. *

An overview of Earthquake design requirements at Pacific Bell is presented. Seismic design considerations for the telephone system including Outside Plant and Telephone Equipment are stated with emphasis placed on the requirements incorporated in the design of telephone equipment buildings.

INTRODUCTION

Seismic design for telecommunications is unique and challenging. At Pacific Bell the need for seismic design is supported and in fact mandated on a corporate level. We recognize our committment to provide service to the public under abnormal conditions as well as the economics of lost service. The nature of the telecommunications system has been discussed at length in other literature (Ref 1) and therefore will not be elaborated upon herein. Briefly the system consists of individual customers connected to a local equipment building known as a Central Office. These central offices are then interconnected through a network of cable, microwave, and toll facilities. At toll offices lines are grouped into trunk routes which interconnect larger population centers.

The most vulnerable part of the telephone system is the switching centers, the Central Offices. Thus most of the attention for seismic design has been given to the central office building itself and the equipment contained therein.

OUTSIDE PLANT

The portion of the system connecting the equipment buildings is generally referred to as the Outside Plant. The outside plant is inherantly earthquake resistant due to its dispersed and redundant nature. The heavily damaged area associated with an earthquake is relatively small in comparison with the size of the telephone network. Thus it

*Project Engineer, Pacific Bell, 3707 Kings Way A2A, Sacramento, California, 95821.

is expected that an earthquake will affect only a small segment of the grid, as has been the case in the past, and due to the redundancy of the grid we expect to be capable of bypassing damaged areas.

There are two considerations which further enhance the seismic resistance of the outside plant. First, a section of slack cable is normally left where the cables enter structures, be it manholes or equipment buildings. This is done to facilitate splicing operations, but it has the positive side effect of permitting relative motion along cable runs without overstressing and pulling apart the cables. There is no mechanism short of a direct rupture of the cable duct by which seismic motion will damage a slack cable and areas of severe ground rupture are relatively small, although such areas may be widely scattered in a large earthquake.

Second, in the past the network was backed up by the Bell System which had the capability to martial enormous amounts of materials and manpower to damaged areas. This benifitted both Bell and non-Bell areas as was demonstrated in San Fernando in 1971 and on numerous other occasions. However the United States Department of Justice has recently eliminated this organization and it remains to be seen if the restructured and separated telecommunications industry can respond to such situations.

There is an additional condideration, one which may decrease the seismic resistance or rebound capability of the system. That consideration is the reduction in the redundancy of the system as mentioned earlier. For various reasons the network is being increasingly centralized. With the increase in centralization the loss of a single facility becomes increasingly a more serious threat to the continued operation of the system.

EQUIPMENT

Telephone equipment has presented a unique problem due to the reluctance of the manufacturers to include realistic seismic considerations in the design of their equipment. Although seismic design is fairly widely accepted in Japan and the United States' west coast, such is not the case in less seismic areas. Thus the manufacturers have been reluctant to add to their products any cost which is requested by a relatively small portion of their customers. This is not to say that adding lateral capabilities in those areas would not be a good idea, it is. This situation has resulted in the need for seismic qualification of equipment, a topic which is being studied by TCLEE (Technical Council on LifeLine Earthquake Engineering).

Pacific braces its equipment to the building structure. The amount of bracing is currently determined by the seismic zone in which the equipment is to be erected based on the Bell System Earthquake Zone Map. This map was in use prior to the breakup of the Bell System and remains in use by Pacific Bell. It closely resembles the Zone Map in the 1982 edition of the Uniform Building Code and utilizes the same zone numbers, 0 thru 4.

The bracing for each piece of equipment is specified in the Bell System Practice number 800-610-155, also still in use at Pacific. Generally it gives the bracing required based on the type of equipment, its configuration, weight, seismic zone and the floor level upon which it will be located. The bracing detailed in this document consists of angles, flat bars, or rods anchored to the columns, walls or ceiling (Fig. 1). All equipment is securely anchored to the floor. The details for this bracing were developed by Bell Labs based in part on computer studies using both Response Spectrum and Time History analysis. Based upon these results suspect equipment, along with items identified by visual inspection as being deficient, were physically tested on a shake table to determine their response and bracing needs.

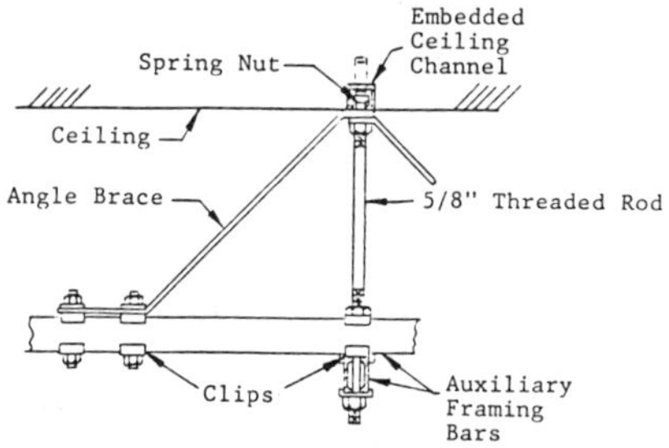

FIGURE 1 - TYPICAL OVERHEAD EQUIPMENT BRACING

This overhead bracing system is most commonly secured to the ceiling above by means of embedded channels (Fig 1). These channels have been routinely installed in new equipment buildings and additions for many years. Where channels are not present self drilling anchors are now used. In the past expansion anchors were used but these have been found in actual use to be less reliable.

A new bracing system has been introduced in the last three years. It is known as the Horizontal Earthquake Bracing System (HEBS) and consists of steel cables strung between the building columns and secured to the tops of the equipment frames (Fig 2). Due to its rubber band appearance it met with initial resistance from the structural community and was affectionately dubbed the "Twang" system. It has since been installed on actual equipment lineups and this experience combined with limited test results indicate good performance may be expected from the system. Its primary advantage is as a retrofitting system in existing, crowded equipment rooms. It is required that prior to installing this system that the building columns be checked for carrying the additional loads.

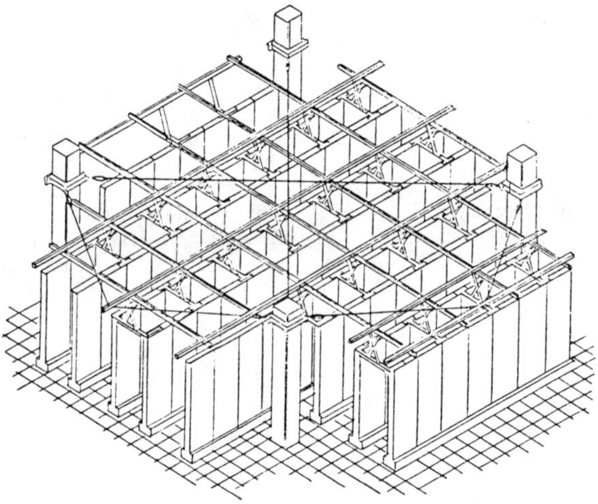

FIGURE 2 - HORIZONTAL EARTHQUAKE BRACING SYSTEM (HEBS)

A final area of concern is for equipment located on raised computer floors. Actual failures of these floors have occured in past earthquakes. The fundamental seismic consideration in a computer floor is to keep the pedestal heights as short as possible. Secondly the system should provide for a means to securely connect the panels and pedestals together and to the structural floor below.

We use a system developed at Bell Labs (Ref 2) using a snubber rod anchorage to secure equipment cabinets located on raised floors (Fig 3). The snubber rod anchorage is a 1/2 inch diameter threaded bar that attaches to the base of the equipment cabinet and then extends downward thru a hole in the raised floor panel to be anchored to the structural floor. The configuration of the anchors varies with the configuration of the particular piece of equipment. The toggle bar is slightly preloaded at installation. This results in the cabinet system acting as a clamp and places the local area of the raised floor in a state of compression, forcing it to act as a three-dimensional frame. Also, the prestress vertical load increases the friction between the feet of the cabinet and the raised floor. Lateral movement of the cabinet relative to the floor will normally occur once the previously described friction forces are overcome. The toggle bar will not resist the initial horizontal equipment motion, but, as the bar is angled loading in the bar occurs and ultimately reaches a level sufficient to completely restrain the equipment from further motion. The toggle bar at all times resists vertical motion and overturning such that hammering of the equipment against the floor is prevented.

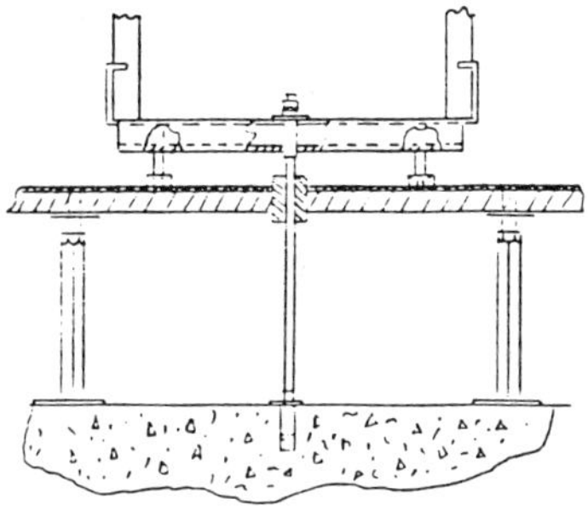

FIGURE 3 - SNUBBER ROD ANCHORAGE FOR EQUIPMENT ON A RAISED COMPUTER FLOOR

This system has been physically tested on the shake table at Bell Labs and found to restrain horizontal movement to approximately 1.5 inches. No visible lifting of the cabinets with respect to the raised floor was observed.

EQUIPMENT BUILDINGS

Telephone equipment buildings are unique structures. Typically they have tall story heights, heavy floor loadings, and provisions for future building growth. For the system to continue operating these buildings must remain in a usuable condition during and following a seismic event. This requires sound and thorough structural design. The structure must be adequately engineered and TIED TOGETHER such that the entire structure WILL ACTUALLY PERFORM WELL, not just meet code requirements.

Our floor loadings are a major difference from typical commercial buildings. The primary consideration is that the given loads are actual loadings and ARE NOT REDUCED FOR AREA as allowed in national codes. These given loads include equipment loads on the floor and ceiling loads beneath the floor in the form of cable racks suspended from the ceiling. Often times roofs are designed as future floors. Where they aren't a live load of 50 psf is used unless Building Code, snow, or actual loads indicate that a greater load should be used. This is done to allow for the addition of unforseen equipment or other future needs.

The individual pieces of equipment are fairly resistant to acceleration, however due to the countless interconnections they are intolerant of substantial relative movements. Therefore in an equipment building deflections are critical and must be kept to a minimum. Since equipment buildings require minimum or no window openings, exterior walls are usually designed as concrete shear walls thereby providing a rigid structure. The use of moment resisting space frames is discouraged. These utilize "ductile" concrete or steel frames which are more flexible and result in higher deflections. Additionally concrete frames must sustain substantial cracking and spalling in order to function as designed. This results in an environment that is not conducive to the continued operation of telephone equipment. Floor deflections are reduced by requiring that thicker slabs and heavier beams be used. Thicker slabs also facilitate the anchorage of equipment from above and below.

Due to the very high cost of relocating an equipment building it is vital that these buildings be designed for future needs. In order to accomplish this future loads and future code requirements must be anticipated in the design. This usually has the positive side effect of producing a structure with further enhanced seismic capabilities.

The typical equipment building is designed for a floor loading of 150 psf and an additive column load, in addition to design loads, of 20 kips per story which is added at the top of all of the columns. Also when considering seismic forces the maximum tensile force is not reduced by more than 50% of the direct stress due to the dead loads. This is in recognition of the effects of vertical accelerations which may occur simultaneously with the lateral accelerations.

In addition to the preceding general considerations the following paragraphs elaborate on some of the particular considerations of importance to the seismic design of an equipment building.

Diaphragms:

The floor and roof slabs of buildings act as horizontal diaphragms distributing the seismic forces to the resisting elements and tying the building together at each level. Due to the numerous cable slots and other openings that penetrate the floors of equipment buildings diaphragms need special consideration in these structures (Fig 4), careful attention to the use of trim bars should be given. All openings and discontinuities must have trim bars of adequate size and length to accomodate not only the usual loads and temperature and shrinkage stresses, but also the diaphragm shears or chord stresses. Included among these should be drag bars to transfer the diaphragm loads to the resisting elements (Fig 5). The anchorage of these bars as determined by bond stresses are not sufficient and must be increased.

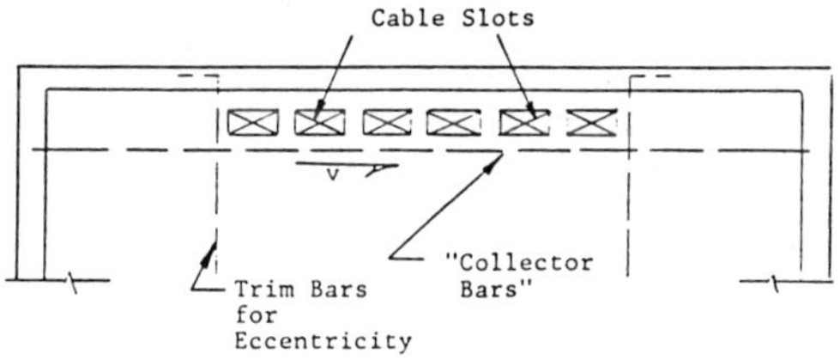

FIGURE 4 - TRIM BARS REQUIRED AT FLOOR PENETRATIONS

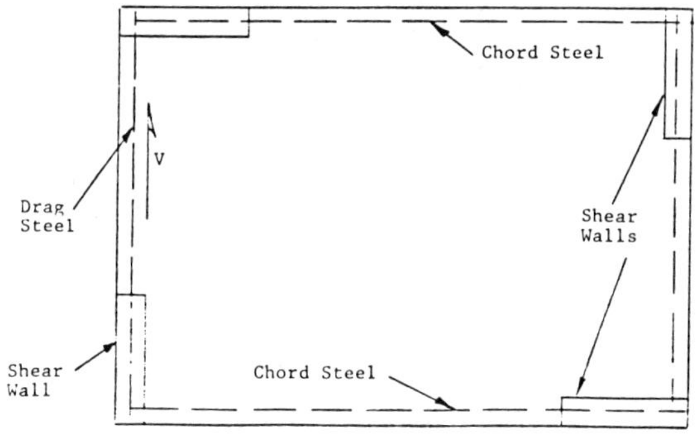

FIGURE 5 - EXAMPLE OF CHORD STEEL AND DRAG BARS

The tensile forces of the slab diaphragm should be carried in the chord bars at the edge of the diaphragm, NOT in the typical slab reinforcing. If carried in slab bars, future cable holes to be cut thru the slabs would create serious problems. Also, in future extensions for growth of the building the larger concentrated chord bars give something substantial with which to tie the old and new portions of the structure together.

From observation of the effects of past earthquakes, the most important single element of earthquake resistant design is the tying together of all elements of the building. Adequate ties to resisting elements must be provided without introducing eccentricities.

Very substantial ties are used between the different units of our buildings. It is difficult to specify an arbitrary amount to cover all conditions, but we require that the minimum tie shoud be 10% of the lighter element being tied. That is, the smaller or lighter portion should be tied to the larger or heavier portion at every slab level with enough steel to resist at least .10 times the weight of the lighter unit. If such connections are impractical, then full seismic separation is required.

It should be noted here that when designing an addition, the addition should be designed for compatibility with the existing structure. The addition should respond similarly to the rest of the structure and should not be substantially more or less flexible.

Torsion:

The effects of torsion on buildings subjected to major earthquake motion has often been observed. Current code requirements do not appear sufficient and the potential effects of torsion are closely examined in our structural designs.

Ideally the resisting elements in a building should be uniformly distributed. However equipment buildings present a unique twist to this problem. Due to the need to permit future growth our buildings often have one or more walls which are intended to be removed in the future (Fig 6). Placing a structural shear wall in these locations is expedient structurally but poses difficult construction problems when the time comes to remove the wall. Light infill walls, such as steel studs and stucco, solve the construction problem but can result in serious torsion problems.

A variety of solutions have been used, none of which provide a perfect solution for all cases. These include grouted block walls, brick, precast panels, steel studs and

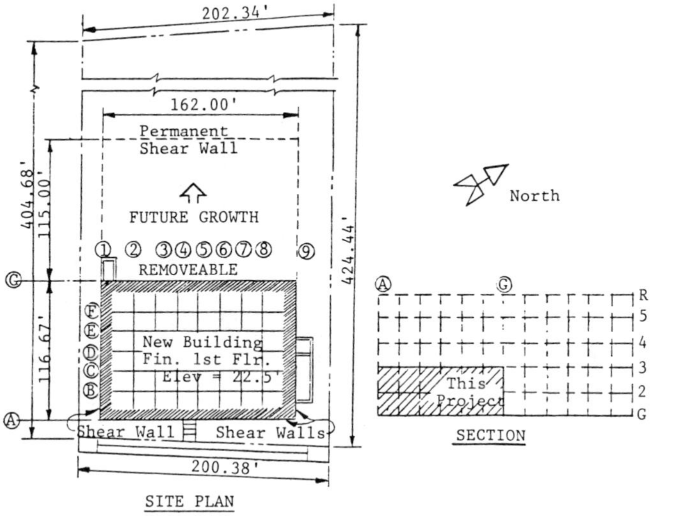

DESIGN CRITERIA:

1. DESIGN LIVE LOADS: This building is designed for the following live loads.

AREA	SLABS & BEAMS	COLS. & FNDTNS.
Typical first floor & roof	150	75
First floor between lines 1 & 2	175	100
First floor between lines D to G & 7 to 9	800	450

2. FUTURE EXTENSION: The walls and columns of this structure are designed to support three additional floors and a roof. Construction similar to the present roof. Building may be extended 115' west of "G" to ultimate permanent shear wall.
3. LATERAL FORCE DESIGN: This building is designed in accordance with the provisions of the 1976 UBC for five floors plus roof.
   V=ZIKCSW, where V=total base shear, Z=1.00, K=1.33, CS=0.14
   W=total dead load plus 70 lbs. psf. equip. load, I= 1.5

FIGURE 6 - EXAMPLE OF DESIGN INFORMATION REQUIRED ON SHEET S1

stucco, and concrete. Whichever solution is used the structural designer and the architect must work closely together to account for the resultant structural response characteristics.

Shear Transfer:

A great deal has been made of insuring that the earthquake loads can be transferred from the structure into the ground. This thinking is backwards. The loads come from the ground and into the structure. Exhaustive efforts to tie the structure into the ground only result in guaranteeing that the structure will receive the full impact of the load.

However, it is essential that the foundations be capable of receiving the load uniformly from the soil. The foundations and the building must be so tied together that they act as a unit. In the typical equipment building for example, if lateral resistance is derived from shear walls, as it often is, then the shear wall forces have to be distributed to all the column footings, side walls, etc., through tie beams and the ground slab so that the shear walls are not displaced relative to the rest of the building. Often this means that the slab on ground is stressed more highly as a diaphragm than the upper floors of the building.

Structural Steel:

Steel is not the preferred material due to the increased deflections often associated with its use. However situations do arise where its use is appropriate. It has most commonly been used for vertical additions where the existing structure would be overloaded by the additional weight of a concrete addition.

When steel is used it is in the form of a braced frame since, as mentioned earlier, the high deflections associated with a moment resisting frame can not be tolerated. The connections for the braces are detailed to develop the full strength of the section thereby eliminating a reduced section which may fail before the full plastic strength of the brace is achieved.

Masonry:

Masonry is generally used by us only for small equipment buildings and administrative buildings. When used it should be fully grouted using a grout of the type that reduces early water loss to the block and produces an expansive action in the grout sufficient to offset initial shrinkage and promote binding of the grout to all interior surfaces of the masonry units. If omitted, the grout will shrink and pull away from the block or brick surfaces, resulting in a wall with questionable capacities.

Wood:

We only use wood in minor structures due to self imposed limitations. Because of our concern for fire safety it is required that in most instances all construction shall be non-combustible.

When wood is used the same concerns outlined earlier for diaphragms are considered. Also wall bracing is required to consist of plywood sheathing acting as a diaphragm. The use of let-in bracing, diagonal sheating, and considering gypsum board or Portland Cement plaster in the lateral force resisting system is not permitted.

Concrete:

Reinforced concrete is the most widely used construction material in our equipment buildings. Its performance can be excellent if a proper system is selected and sound detailing of the reinforcing steel is incorporated into the design. It is also important, as mentioned earlier, that functional flexibility be provided in the design. This means that the structure should have the ability to support extra loads, have holes or slots cut in the floors, etc.

The most commonly used structural system is a concrete frame incorporating concrete shear walls for lateral support. Floor slabs are required to have a minimim thickness of 5 inches, however 8 inches is commonly used.

Seismic considerations and the required flexibility to alter the structure require careful attention to details. Creep and subsequent loss of camber must be considered in order to insure flat floors. Thus reinforcing is required to be continuous throughout the slabs and beams, both top and bottom.

Some minimum requirements are given to our designers regarding the reinforcing details of our buildings. These include the use of a minimum positive and negative reinforcement of .005bd. Beams and girders are to have 25% of the positive and negative reinforcement (2-# 5 minimum) continuous top and bottom with 40 diameter laps. At exterior spans, bars are to be anchored with 90 degree hooks with continuous steel top and bottom equal to .0025bd each. Stirrups are required throughout beams spaced d/2 at the ends (1/6 span) then 18" throughout.

Also 25% of joists are required to have a #5 bar continuous top and bottom and slabs are required to have 25% of the positive and negative reinforcement continuous throughout the slab, lapped and anchored at the exterior of the slab.

Shear walls are commonly used but when used alone are limited to a height to length ratio of 3:1, or a height of 160 feet, whichever is less. Higher walls are permitted when combined with a moment resisting frame capable of resisting at least 25% of the lateral forces.

When cantilever shear walls are used the moment resisting steel is concentrated at the outside. This increases the ductility - up to double the case where it is distributed over the length of the wall. When this is done all bars near the ends of the wall, the flanges, are tied to prevent buckling. The tie spacing is limited to six bar diameters. Also enough horizontal steel is required to prevent shear failure and to permit the developement of hinging.

A final item of interest with regards to concrete is a new process being researched by others which uses fiber reinforcing in concrete joints. This could be very valuable in seismic design due to the greatly improved ductility demonstrated in early tests. It also offers additonal promise for equipment buildings since it appears to greatly reduce spalling which could improve the post-earthquake equipment environment in addition to improving the structural response. Fiber reinforcement is not used by us but may be worthy of future consideration.

Columns:

Column design is critical to the performance of a concrete building. As mentioned earlier we require an additional 20 kip load per story to insure good performance and permit future growth. The requirement that live loads not be reduced further reinforces this need. In addition to this we desire that our buildings not be obsoleted by future code changes.

In order to insure good performance we require a maximum reinforcement ratio of 4% and bar laps of 40 diameters for grade 40 rebar. Also all bars must be tied and the ties are a minimum of 3/8 inches diameter with a maximum spacing of 12 inches. At beam/column joints ties are carried thru the joint and 24 inches beyond with the ties spaced 4 inches on center. Although not currently required by our standards (Ref 3) a closer spacing of ties then required by the codes should be considered for the full height of the columns.

Column splices are not permitted in the joints and in seismic zones 3 and 4 all columns are designed to resist an additional lateral earthquake load 7 to 9 feet above the floor level. This requirement has been recently added in order to account for the cable bracing system discussed earlier.

Towers:

The question of insuring the performance of microwave (and transmission) towers has often been raised. However, due to their relatively light weight construction and long period they do not have a substantial problem with earthquake loading. In the design of towers wind, not earthquake, usually governs.

Documentation:

In order to assure good performance of a structure the engineer must have accurate design information. This is often a problem when adding to an existing building and the structural engineer must guess at what the original loads, criteria and future growth plans were. We avoid this problem by requiring that the basic design information be included on the first sheet of the structural drawings. An example of this information is illustrated by Figure 6.

BUILDING SUPPORT EQUIPMENT

Mechanical, electrical and certain other non-telephone systems in equipment buildings are designed and braced with the intent that they remain operational after an earthquake. This requires anchoring or restraining each piece of equipment to prevent sliding, overturning or hammering. Sensitive equipment must have an internal design that will remain functional and the equipment frame must be compatible with external bracing.

All support equipment is designed with local codes as a minimum and also must meet the provisions of the California State Hospital Code where more stringent. Additionally we specify minimum lateral loads for various pieces of equipment. Rigidly mounted equipment necessary to the continued operation of the facility is required to withstand a lateral load equal to 75% its own weight acting simultaneously with a 33% vertical load. Flexibly mounted equipment must handle 100% lateral and 33% vertical. Piping, conduit, air ducts, etc. must be designed for 50% lateral except that smaller sizes are exempted.

It is vital to the continued operation of these facilties that their power supply be maintained under all conditions, seismic events included. The commercial power facilities within the building are protected by the steps just outlined. However, the security of the power system beyond the House Service panel is beyond our control and it has been found that commercial power cannot be relied upon. Therefore, all but the smallest equipment buildings are provided with a Standby power system which typically consists of a battery plant which picks up the load immediately upon the loss of commercial service. This is then supplemented by either portable or stationary generators which maintain the charge on the batteries. Stationary generators are generally replacing the portable generators relied upon in the past because it has been realized that following a catastrophic event transporting portable generators into the stricken area may be difficult or impossible. It is required that stationary generators meet the design criteria outlined above.

ECONOMICS

Taken as a whole the requirements outlined herein may seem excessive. However, the actual cost of these precautions must be weighed against their worth.

Over the years these standards have been applied to hundreds of new buildings and additions and we find that they add approximately 10% to the construction costs. This is significant when considering a commercial building. However, the overall project cost must be considered. For commercial buildings the building typically constitutes 90% of the cost and its contents 10%, thus these standards would add 9% to the total project cost.

In the case of telephone equipment buildings these numbers change radically. The equipment that goes into these buildings costs on the order of 10 times as much as the building. Thus the building is 10% of the project cost and the equipment 90%. Thus these standards add only 1% to the total project cost.

Putting telephone equipment in a cheap building is analogous to shipping computers in paper bags. A bag is appropriate for potato chips but not for expensive hardware.

CONCLUSIONS

Most of the practices outlined herein have been in use for several years. Telecommunications facilities have performed well in past seismic events although this must be partially attributed to luck. In recent years there have been only two seismic events which substantially shook telephone facilties.

The first was the San Fernando Earthquake in 1971. That earthquake inflicted major damage to another company's central office which was located in the area of intense shaking. In that case the building was designed in a manner similar to that described herein and suffered little damage. However, the equipment inside was inadequately braced and was destroyed.

The second was the Coalinga Earthquake in May 1983. Pacific Bell's central office building in Coalinga was built long before these standards were formalized, however, it was designed by Henry Degenkolb who was instrumental in the formulation of these standards. It is evident in the design of that building that Mr. Degenkolb employed the philosophy behind these standards in the design and his foresight paid off. Despite heavy damage to most of the surrounding buildings our building sustained very minor damage, primarily non-structural. The main area of damage was confined to the growth wall at the rear of the building, a situation discussed earlier, which sustained considerable cracking.

The equipment within the building withstood the earthquake and numerous aftershocks with only minor damage to battery racks and some signs of initial yielding in overhead bracing. The system continued to function with service interruption confined to the limitting of incomming calls. This is done by the system during such an occurence in order to prevent saturation of the local network. Based on this experience our efforts in seismic design appear to be successful and very much worth the cost and effort expended.

ACKNOWLEDGEMENTS

I must first acknowledge the management of Pacific Bell, particularly our District Manager, D. S. Cannon, in permitting this paper to be written. Their support and the access to information which they provided were invaluable.

The information on Outside Plant and Equipment was supplemented by my fellow employees including Ben Briones, Jim Smith and Nick Papson. John Foss of Bell Communications Research, formally of Bell Laboratories, provided most of the information on the design of the equipment bracing.

The buildings standards set forth herein are contained in Pacific Bell's Structural Design Standards (Ref 3) which I revised and editted based on an earlier document. That document was originally written by Henry J. Degenkolb, J. Albert Paquette, and Robert E. Jones and their original work is still very much evident in this paper.

REFERENCES

1. Advisory Notes on Lifeline Earthquake Engineering, Technical Council on Lifeline Earthquake Engineering, American Society of Civil Engineers, New York, New York, 1983, Library of Congress number 83-72329.

2. J.W. Foss and G. A. Nikolakopoulou, "Electronic Data Processing Equipment Braced to Resist Seismic Loading", Conference Proceedings, Seventh World Conference on Earthquake Engineering, Istanbul, Turkey, September 1980.

3. Structural Design Standards, January 1, 1984, Pacific Bell, San Francisco, California, 94105.

Seismic Risk and Reliability of Telecommunications Systems

S.C. Liu* and J.N. Yang**

A general method for seismic risk analysis of telecommunications systems is developed. The method evaluates by statistical means the earthquake environment and the earthquake vulnerability and effects on the system under consideration. Methods to assess the structural reliability of these systems are presented and concepts of cost-optimum design to protect these systems against earthquake damages or failures are introduced. Examples of analysis of community dial office and the equipment supported therein is presented.

Introduction

While it is essential to maintain their operations during and after earthquakes, telecommunications systems are vulnerable to earthquake damages. In the conventional earthquake design and analysis, the uncertainty due to the randomness in earthquake occurrences and the variability in the induced ground motion are generally not accounted for. As a result, the earthquake risk and structural loads may be under- or overestimated, and hence the structure or system may be improperly designed. An underdesigned telecommunications structure is clearly unacceptable because excessive monetary loss and structural damage may be sustained during its service life. On the other hand, an overdesigned structure results in an unnecessary expenditure for excessive earthquake protection that is not commensurate with the risk involved. This, coupled with the continued demand for reliable, yet cost-effective engineering, has enhanced the importance of the search for optimal structural designs, which include both the protection cost and the monetary value of the expected earthquake loss as key design factors.

Proper structural design, therefore, must include consideration of both cost and risk in order to insure that design decisions maintain a balance between the total cost and reliability of a structure in a specified earthquake environment.

This total system investment is comprised of earthquake damage costs (including loss of services) and construction costs for building, equipment, and distribution plants (i.e., cables, microwave radio) that meet prescribed levels of earthquake protection. Within the context of cost-effective seismic design, as opposed to conventional practices, the

*Adjunct Professor, The George Washington University, currently with National Science Foundation, Washington, DC 20550.
**Professor, The George Washington University, Washington, DC 20052.

major concerns for a communications facility are: (1) to express the earthquake hazard as a failure probability in terms of the earthquake magnitude and the occurrence statistics, (2) to correlate the earthquake damage and interruption of service with a monetary loss, and (3) to make design decisions that insure a balance between cost and risk.

The objective of this paper is to develop a general procedure for cost-effective design that can be applied to telecommunications building-equipment systems. This procedure is applied to the analysis of Community Dial Offices for illustration of the methodology.

Analysis Procedure

The schematic diagram in Fig. 1 illustrates the concepts involved in a closed-loop procedure for designing a cost-effective building-equipment system for operation in a random earthquake environment. Basically, two independent sources of information that define the earthquake environment and the corresponding structural response are considered. The information that pertains to the earthquake includes the geological and historical earthqauke data and is used to describe the shock-induced ground motion. This data is usually specified by the probability distribution for the earthquake Richter magnitude, the regional attenuation law, and the source geometry. Ground motion statistics can be presented (see Reference 1) in which the probability distributions for the site intensity and the peak ground acceleration are derived.

The structural information, on the other hand, quantifies the material properties that pertain to the foundation-structure-equipment system. Preliminary studies based on this data are used to identify the potential failure modes of the structure. The earthquake-structure interaction is then defined by the response to earthquake ground motion in terms of these failure modes. In order to specify the failure probability, the likelihood that the response exceeds the material limits of the supporting members must be determined. Furthermore, when these limits are exceeded, a damage cost associated with the structural failure is given or assigned.

Failure and Loss Analysis

The basic assumptions made are that the structure under consideration belongs to the class of linear, time invariant systems, and that all the structural damage is defined in terms of a given set of independent failure mode responses. In particular, it is assumed that for each failure mode there is an allowable range of response values. Their combined effect is to define a domain of safe response for the structure as a whole.

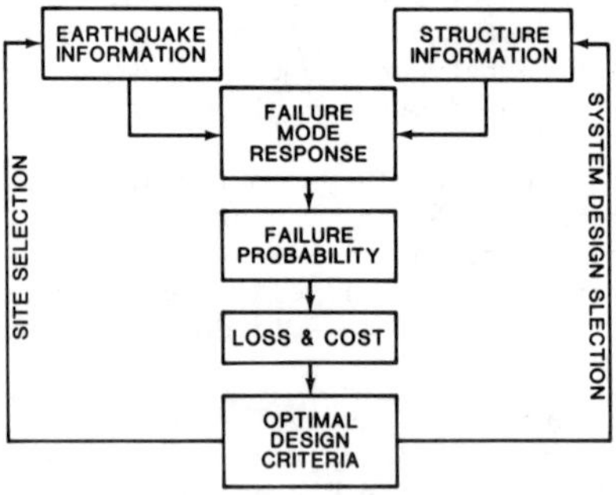

Fig. 1 Cost-Effective Seismic Design Approach.

Letting $r_j(t)$ be the response time history of the jth failure mode to the earthquake ground acceleration of duration T_0, our attention is focused on the following types of allowable response domain:

$$0 \leq r_j \equiv \sup_{t\varepsilon[0,T_0]} r_j(t) \leq R_j \tag{1}$$

$$0 \leq r_j \equiv \sup_{t\varepsilon[0,T_0]} |r_j(t)| \leq R_j \quad , \tag{2}$$

Equations (1) and (2) are referred to as one-sided and two-sided domains, respectively, because of the distinction between the allowable response excursions--the excursion of the maximum positive response is restricted in the one-sided domain; whereas, the maximum excursion in both directions is restricted in the two-sided domain.

The failure mode response, $r_j(t)$, to the earthquake ground acceleration is defined by the associated transfer function, $H_j(\omega)$. Using the zero mean, stationary, and normal acceleration process for earthquakes, the inherent linearity of the structure implies that similar attributes apply to the response process. Furthermore, the convolution theorem implies that the compounded effect of the linear ground filter and the earthquake-structure interaction is prescribed by the product of the separate transfer functions $H_g(\omega)$ and $H_j(\omega)$, where $H_g(\omega)$ is the ground filter transfer function characterizing the earthquake acceleration process. The peak failure mode response is, therefore, amenable to the analysis of linear system response to stationary random vibration, and the resulting relations are

$$\sigma_{r_j} = H_{0j} S_n^{1/2} \tag{3}$$

$$H_{kj} = \left[\int_{-\infty}^{-\infty} |H_g(\omega) H_j(\omega)|^2 \omega^{2k} d\omega \right]^{1/2} \quad , \quad k=0,1,\ldots \quad . \tag{4}$$

where S_n is the constant power spectral density value characterizes the intensity of the ground accelerations. With the assumption that the number of zero crossings remains relatively large, the peak response, given the event, E_i, that an earthquake site intensity I is equal to i, is approximated by the Gumble distribution

$$P[r_j \leq r | E_i] = \exp\{-\exp\{-\alpha_j (r-\mu_j)\}\} \tag{5}$$

$$\alpha_j = (2 \ln \nu_j T_0)^{1/2} / \sigma_{r_j} \tag{6}$$

$$\mu_j = \sigma_{r_j} (2 \ln \nu_j T_0)^{1/2} \quad . \tag{7}$$

The distribution between one-sided and two-sided allowable response domains is apparent in the expression for the average zero crossing rate, ν_j

$$\nu_j = H_{1j} / 2\pi H_{0j} \qquad \text{(one-sided)} \tag{8}$$

$$\nu_j = H_{1j} / \pi H_{0j} \qquad \text{(two-sided)} \tag{9}$$

Consequently, this distinction is also implied in the scale and location parameters, α_j and μ_j, respectively.

The assumption that the intensity of the earthquake is not dependent of the occurrence time implies that the response characteristics are time invariant. Therefore, the occurrence of failures during the service life of the structure is properly described as a homogeneous Poisson process.

The conditional probability that the peak response of the jth failure mode, r_j, is outside the range of allowable response values, R_j, given an earthquake of site intensity I = i has occurred is obtained from Eq. (5)

$$P_{F_j|I} = P[r_j > R_j | E_i] = 1 - \exp\{-\exp\{-\alpha_j(R_j - \mu_j)\}\} \; . \tag{10}$$

Due to the independence of the failure modes, the probabilities given by Eq. (10) can be combined to form the conditional failure probability for the system as a whole

$$P_{F|I} = 1 - \prod_j \left(1 - P_{F_j|I}\right) \; . \tag{11}$$

The conditioning variable I in Eq. (11) can be removed by using the site intensity probability p_I

$$P_{F|E} = \sum_I P_{F|I} \, p_I \tag{12}$$

Thus, with N_F denoting the number of failures that occur during the service life of the structure, the failure probability, $P[N_F > 0]$, can be used to evaluate the structural design from a reliability point of view.

The crucial step in the assignment of damage costs is to correlate the dollar value of the earthquake loss with the failure mode response to the earthquake ground acceleration. For any earthquake it is assumed that the loss is a simple random variable that has the value zero in the event a failure does not occur, and has the value, ℓ, in the event a failure does occur. Since the occurrence of failures is described as a homogeneous Poisson process, the average and variance of the earthquake loss are

$$\overline{L} = \overline{N} \, \ell \tag{13}$$

$$\sigma_L^2 = \overline{N} \, \ell^2 \tag{14}$$

where $\overline{N} = \alpha T \, P_{F|E}$ is the expected number of failures during the time period, T, α is the average occurence rate of earthquakes and $P_{F|E}$ is the conditional probability that an earthquake occurrence results in a failure.

Because of the randomness inherent in the occurrence of structural failures, a precise dollar value cannot be placed on the earthquake loss incurred during the service life of the structure. Instead, the average (expected) loss associated with all possible failure configurations is used as a measure of the earthquake damage cost. This cost depends on the failure probability as prescribed in Eqs. (11) and (12). Since the failure probability depends on the level of protection designed into the structure, the protection cost can be compared with the damage cost to determine an optimal level of protection investment.

The criteria used to specify the "optimal level" could include, for example, constraints that limit the failure probability and the ratio of protection to damage costs. For approximation, the total cost is defined as the sum of the protection and damage costs. By comparing the total cost for all practical design alternatives the optimal design can be identified as that associated with the smallest value of total cost.

Analysis of Community Dial Office in an Earthquake Region

The method of seismic risk analysis and the cost-effective earthquake design procedure will be applied in this section to telephone communications structures. In particular, a typical Community Dial Office (Ref. 2) in an earthquake susceptible area like California is considered for analysis.

Standard CDO office services as a collection and distribution point for local phone users with a size ranging from 400 to 500 lines, although there exist nonstandard offices of over 5000 lines. Analyses and tests seem to indicate that several deficiencies are inherent in the present specifications that outline the protective measures currently used in CDO systems. A typical CDO support system is shown in Fig. 2. The support structure ties the equipment frame lineups together at the 9 foot level by means of cross-aisle cable racks and 1-1/2 x 1-12 x 3/16 inch angle braces (referred to as lineup braces in Fig. 2). This superstructure is anchored by wall braces attached to wood battens, which in turn are secured to the wall by 3/8 inch lag screws in frame buildings, or toggle bolts in concrete buildings. The Common Distribution Frame (CDF) and Opposite side (OPP) wall braces have been identified as critical components, and the primary failure mode was found to be the brace-batten connection failing in tension. The total wall load, W, in the weak-stiffness direction was assumed to be uniformly distributed over the individual wall braces so that $W = fB$ where B is the number of braces and f is the load per brace. It was found that the dynamic characteristics depend primarily on the equipment weight, that they do not change appreciably even when the number of braces is doubled, and that on the average the total wall load depends linearly on the weight of the equipment in the office. Furthermore, the equipment weight was shown to be roughly proportional to the equivalent length, L, defined as the length of the CDF times four, plus the length of the remaining lineups.

Assuming the CDO office is located in an earthquake environment for which the following characteristic constants are used in the calculations: the Poisson arrival rate, $\alpha = 6.33$

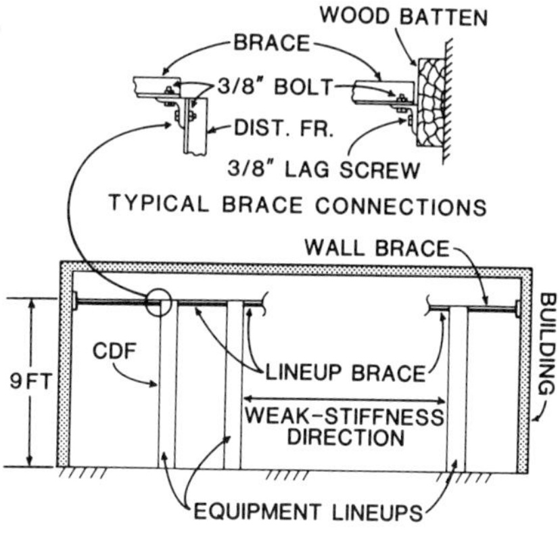

Fig. 2 Typical CDO Support System.

per year, the threshold earthquake magnitude, $m_0 = 4.5$, the regional seismicity constant, $\beta = 0.87 \ln 10$, the focal depth, $h = 20$ km, and the fault length, $l = 644$ km.

Based upon the formulations presented above, the essential statistics describing the earthquake environment can be derived. For example, Fig. 3 shows the probability of the site intensity given that an earthquake has occurred some distance away from the fault. It is clear that this probability depends on the regional magnitude probability, the site-to-fault distance, and the attenuation relation. The major effect of increasing the distance is to increase the probability of smaller intensity events, and to significantly decrease the probability of larger intensity ones. For all distances, however, the probability of the larger events decreases in an exponential fashion with the intensity level.

Further information about earthquake environment can also be obtained by exploiting the formulation presented earlier. Such information may be displayed in terms of probability of the occurrence of one or more earthquakes with site intensity $\geq i$, return period for earthquakes with site intensity $\geq i$, or percentiles of peak ground acceleration for an earthquake with a given site intensity.

The two-dimensional structural model in Fig. 2 with two configurations for the equipment lineup connections, i.e., fixed and pinned, is employed here. For either configuration, the structural model is linear, and thus amenable to a standard modal analysis. The modal damping is assumed to be 2% for all modes. For a typical CDO installation, the critical structural elements are the wall braces, and as mentioned earlier, the primary failure mode is the brace-batten connection shown in Fig. 2 which is assumed to be unable to withstand tensile loads in excess of 900 pounds. Since the structural model is linear and elastic, the wall brace load is directly proportional to the difference between the member node point displacements, y. Letting j_1 and j_2 denote these points for a given wall brace, the stress-strain relation implies that the load per brace, f, is given by

$$f = \left(y_{j_2} - y_{j_1}\right) AE/\ell \tag{15}$$

where each brace is characterized by a cross-sectional area, A, of 0.53 in^2, an elastic modulus, E, of 3×10^7 psi, and a length, ℓ, of 57 inches. Furthermore, from the results of the deterministic study, the total wall load is proportional to L/B.

This proportionality is of tremendous advantage because the results of a single analysis can be directly extended to other designs by simply invoking the scaling property. As an example, consider a hypothetical CDO having an equivalent length of 268 feet, the recommended bracing system according to the standard design consists of 22 braces, 13 on the CDF side and 9 on the OPP side. A brace design factor, ρ, can be defined as

$$\rho = B_s/B \tag{16}$$

where B_s is the number of required braces (standard design) and B is the actual number of braces used in the CDO. As a result, the scaling property is defined by the relation

$$f = \rho f_s \tag{17}$$

where f_s is the load per brace for the standard design.

An important property of the design factor is that it only affects the modal participation factors, since the modal damping, natural frequencies, and mode shapes remain unchanged. The brace design factor is a measure of the protection level in the sense that a design factor smaller (larger) than unity means that more (less) braces than the standard design are used. According to the definition of ρ in Eq. (16), the case $\rho = 1$, may be referred to as the standard design, the case $\rho = 1.33$ as 25% underdesign, and $\rho = 0.8$ as 25% overdesign.

The failure mode response, or load per brace, conforms to the general model developed earlier with a one-sided domain of safe response. Since a brace failure occurs only if the tensile load exceeds the pullout strength of the brace-batten connection, the appropriate failure statistic is the peak tensile load, f_p, in the brace. The conditional peak response given the occurrence of an earthquake with intensity $I = i$ is governed by the extreme value distribution similar to Eq. (5).

The conditional probability of failure, $P_{F|I}$, given that an earthquake with a certain intensity has occurred, is given by Eq. (10) using a pullout strength of 900 pounds. The results are plotted in Fig. 4 as a function MMI. This figure consists of four curves corresponding to various brace design factors, i.e., 50% overdesign ($\rho = 0.67$), 25% overdesign ($\rho = 0.8$), standard design ($\rho = 1.0$), and 25% underdesign ($\rho = 1.33$) with respect to the standard bracing design. From these curves, it is clear that the CDO conditional failure probability increases in a roughly exponential fashion as the intensity level is increased. A clearly unacceptable level of risk must be assumed when using the 25% underdesigned bracing system. In a moderate risk region where many earthquakes with MMI $\geq$ VI are likely to occur during the service life of the CDO structure, the failure probability for a 25% underdesigned CDO is 10%. For the standard and overdesigned bracing systems at the same probability level ($P \simeq 0.1$) the failing intensity values are MMI $\geq$ VIII, $\geq$ X, and $\geq$ XII, respectively. The actual risk involved, however, depends on the site distance because the return periods at the more distant site locations for these higher intensity levels are much larger than for lower intensity levels. Therefore, the actual risk can be said to depend on the earthquake occurrence statistics, and is more properly characterized by the probability $P[N_F > 0]$ that one or more earthquakes will induce a failure of the CDO protection system during the service life of the structure. This probability is presented in Fig. 5 as a function of the site-to-fault distance for the CDO system in the prescribed hypothetic earthquake environment. As auxiliary variables, we consider the service life (25, 50, and 100 years) and the brace design factor (25% underdesign, standard design, 25% overdesign, and 50% overdesign) for twelve possible configurations. In each case, the probability decreases in a roughly exponential manner with epicentral distance, and increases with the service life. These trends, however, are less apparent for the 25% underdesigned protection system because the probability is approaching the asymptotic value of unity. This further supports the earlier contention that an unacceptable level of risk must be assumed when using the 25% underdesigned bracing system. Even at a distance of 100 km, the 25 year failure probability is 0.21, corresponding to a reliability or survival probability of 0.79.

Considerable improvement in the reliability of the protection system is achieved by adopting the standard design. At the more distant site locations, the likelihood of a failure drops to a value between 1 and 10 percent. However, the failure probability for a site located near the fault system is at a level comparable in risk to that for the 25% underdesigned case at a distance of 100 km. Continued improvement in reliability is obtained by employing the 25% and 50% overdesigned protection systems. The failure probability drops by more than an order of magnitude for the 25% overdesign case, and much more for the 50% overdesign case, with respect to the standard design.

Earthquake Damage Costs Analysis

By using the stochastic model, the reliability of a CDO installation for various protection levels, service lives, and site locations can be evaluated. Such information provides the design engineer with a powerful method for discriminating between reliable and unreliable structural designs. Ideally, the engineer would choose the design with the least overall cost that meets the minimum reliability requirements.

To clarify this point, the total cost is defined as the sum of the losses resulting from earthquake occurrences during the service life of the structure, plus the additional costs for providing the earthquake protection system. It is assumed that a loss is incurred only if the peak load in the CDF wall brace exceeds the assumed allowable response limit of 900 pounds, and when this does occur, the loss is a fixed amount equal to the replacement cost of the equipment, which is taken as $1.5K per linear foot of equivalent length of the CDO equipment.

A precise dollar value cannot be placed on the total operating cost during the service life of the structure because of the random occurrence of earthquake induced failures. It is natural in this case to use the average earthquake loss as a measure of the damage cost. The total operating cost is then obtained by adding in the cost of the protection system. In Fig. 6, the earthquake loss given by Eq. (13) is presented as a function of epicentral distance. The effects of the brace design factor and the duration of the service life are identified by distinguishing the results for the 25% underdesign, standard design, 25% overdesign, and 50% overdesign cases for 25, 50 and 100 year service lives. Each curve exhibits the same general trend to increase as the service life increases, and to decrease in a roughly exponential manner as the epicentral distance is increased. Furthermore, as the level of earthquake protection is increased, the average loss is reduced by roughly an order of magnitude for each 25% change in the brace design factor.

A great loss for a CDO employing a 25% underdesigned protection system requires special attention. Because any loss greater than $1.5K per linear foot of equivalent length of CDO equipment implies that more than one failure will be experienced during the service life. Such a high loss is associated with the high failure probability of the underdesigned case as indicated in Fig. 5.

Cost Effective Seismic Design

For comparison purposes, protection costs are included in Fig. 6 based on the two values of \$0.25K and \$1.00K per brace. For standard design cases, these two values correspond to protection cost ratios of 1.4% and 5.6% respectively, where the protection cost ratio δ is defined as the brace cost divided by the equipment cost. These curves appear as straight lines because they are independent of the site-to-fault distance. Taken as a lower and upper bound of the cost per brace, they serve to define a protection cost range that can be compared to the average earthquake loss. For example, the ratio of the protection cost to the earthquake loss can be used as one measure of the cost-effectiveness of design. Using this ratio, we consider a design as cost-effective (but not necessarily reliable) whenever the earthquake loss exceeds the protection cost. In effect, a design is cost-effective provided that the investment in the protection system is not more than the expected loss during the service life of the structure. From Fig. 6, the 25% underdesigned protection system is always cost-effective (but from Fig. 5, unreliable), and the 50% overdesigned one is not cost-effective (but reliable). The intermediate designs are cost-effective for a CDO installation located near the fault, but not cost-effective at the larger distances.

Optimal Protection

It should be emphasized that there is no unique definition for cost-effective design or optimal design. Basically the cost-effectiveness is judged by the comparison of two different cost measures, while the optimization is determined by the minimization or maximization of a certain objective function. A cost effective design is not necessarily optimal, and an optimal design may not be cost-effective. In this analysis an optimal design is considered as the one that minimizes the total cost of the structure, i.e., the combined value of the expected earthquake loss and the protection cost of the structure. In the limiting case where no protection braces are provided ($B \to 0$ in Eq. (16)) there is no protection cost but the equipment loss and the maximum total cost could be the total equipment value. In the other extreme where infinitely high protection is provided ($B \to \infty$), there will be no failure and thus no loss, but the protection cost and hence the total cost is infinite. Somewhere between these two extremes there exists a design for which the total cost is minimum. Such an optimal protection concept can be illustrated by considering the total cost of a typical CDO office in a prescribed seismic region for various brace designs, service lives, and epicentral distances. Figure 7 presents the results in terms of the ratio of the total cost to the total equipment value for a protection ratio $\delta = 2\%$. The total cost increases linearly with the number of braces in the extreme overdesigned limit; while this cost increases exponentially with the reciprocal of the number of braces in the extreme underdesigned limit. Between these limits, there is a unique minimum for the total cost curve and a corresponding optimal brace design level.

From Fig. 7 and for practical purposes, the 20% overdesign case can be considered approximately as the optimal protection level for CDO systems in the seismic region located close to the causativefault.

Conclusions

A general method for seismic risk analysis of telecommunications structures has been developed. The structural model considered is typical of single or multistory telephone buildings with floor-, ceiling-, or wall-supported communications equipment. The method is applied to a CDO system in a seismic region to demonstrate its usefulness.

References

1. Cornell, C.A., "Engineering Seismic Risk Analysis," Bulletin of Seismological Society of America, Vol. 58, No. 5, October 1968, pp. 1583-1606.

2. Isenberg, J., "Seismic Analysis of Telecommunications Equipment-Building Systems," ASME Symposium on Lifeline Earthquake Engineering, Portland, Oregon, June 1983.

3. Liu, S.C., and Neghabat, F., "A Cost Optimization Model for Seismic Design of Structures," BSTJ, Vol. 51, No. 10, December 1972, pp. 2209-2225.

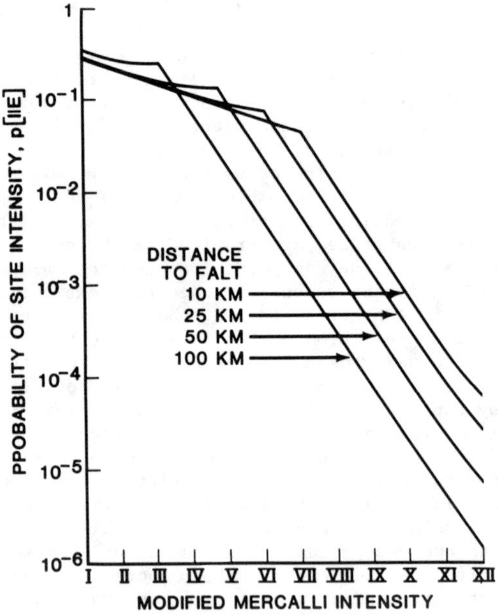

Fig. 3 Probability of Site Intensity Given an Earthquake Occurrence.

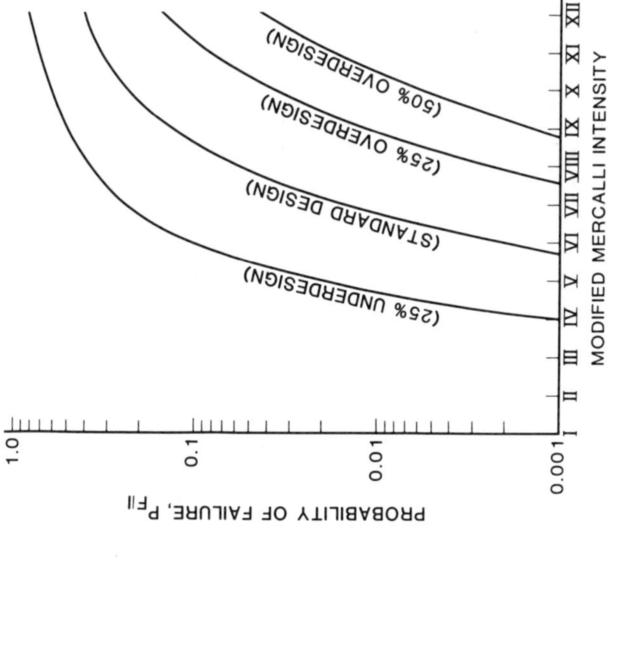

Fig. 4 Conditional Probability of CDO Failure Given an Earthquake Occurrence.

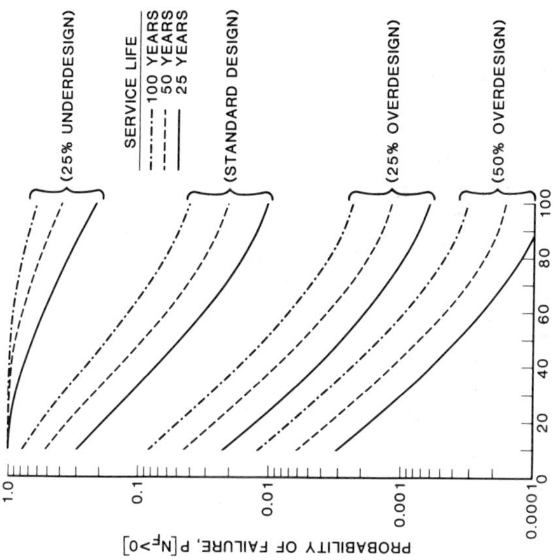

Fig. 5 Failure Probability of CDO.

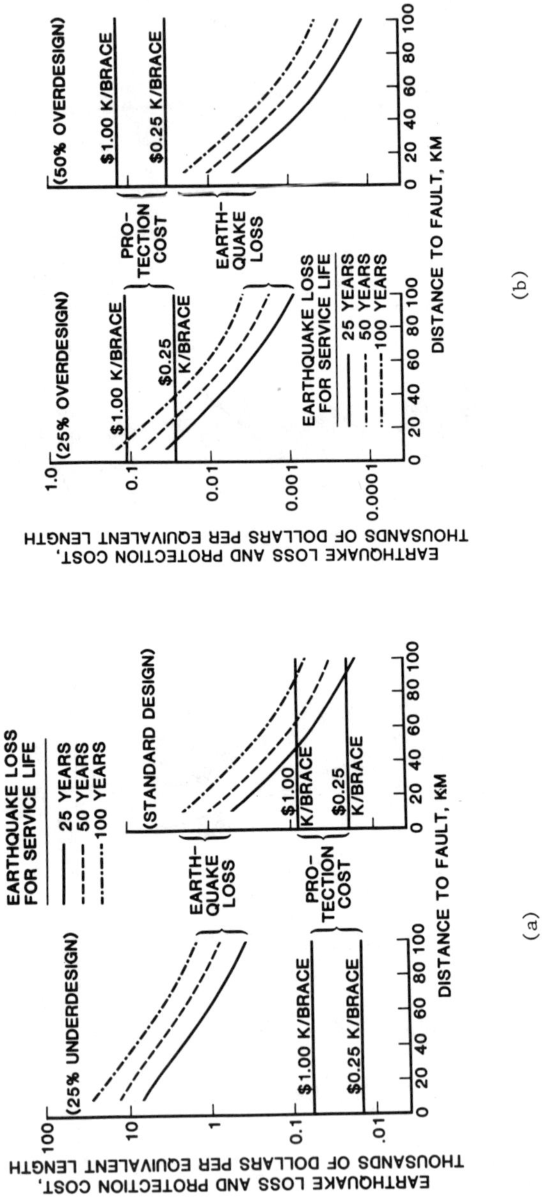

Fig. 6 Earthquake Loss and Protection Cost for CDO.

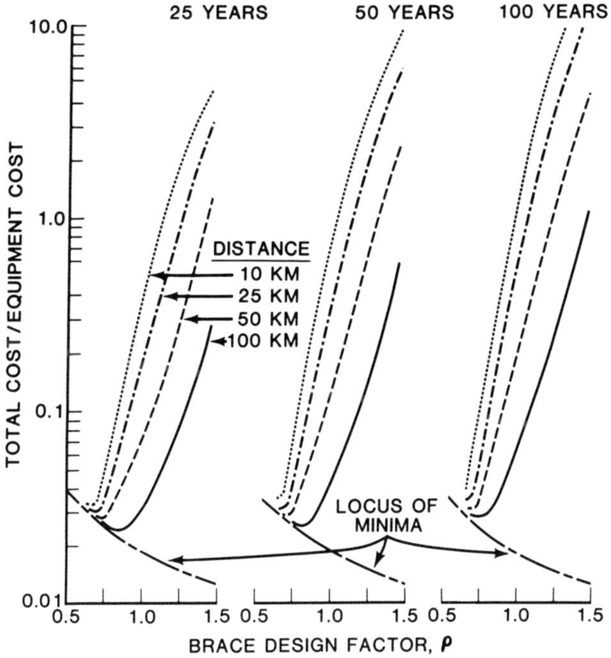

Fig. 7 Cost Ratio For CDO in a Seismic Region For $\delta = 0.02$.

SUBJECT INDEX

Page number refers to first page of paper.

AUTHOR INDEX

Page number refers to first page of paper.